Importance of Oxidative Stress and Antioxidant System in Health and Disease

Edited by Suna Sabuncuoğlu and Ahmet Yalcinkaya

Published in London, United Kingdom

Importance of Oxidative Stress and Antioxidant System in Health and Disease
http://dx.doi.org/10.5772/intechopen.100949
Edited by Suna Sabuncuoğlu and Ahmet Yalcinkaya

Contributors
Xian Zhou, Gerald Münch, Dennis Chang, Emma Borrelli, Ngoc-Anh Le, Bhavana Gangwar, Santosh Kumar, Mahendra P. Darokar, Emina Čolak, Lepša Žorić, Miloš Mirković, Jana Mirković, Ilija Dragojević, Dijana Mirić, Bojana Kisić, Ljubinka Nikolić, Maria G. Barderas, Laura Mourino-Alvarez, Tamara Sastre-Oliva, Nerea Corbacho-Alonso, Livia Rodrigues, Igor Silva, Matheus Santos, Pedro Lirio, Ayodeji Folorunsh Ajayi, Moyinoluwa Comfort Onaolapo, Samuel Chibueze Nzekwe, Lateef Okeleji Olabisi, Victor Oluwaseyi Amos, Oluwatobi Hezekiah Ajayi, Gustavo Alberto de la Riva, Luis Adrian Saldaña Trujillo, Juan Carlos González-Hernández, Olalekan Bukunmi Ogunro, Aderonke Elizabeth Fakayode, Gaber El-Saber Batiha

First published in London, United Kingdom, 2023 by IntechOpen
IntechOpen is the global imprint of INTECHOPEN LIMITED, registered in England and Wales, registration number: 11086078, 5 Princes Gate Court, London, SW7 2QJ, United Kingdom
Printed in Croatia

British Library Cataloguing-in-Publication Data
A catalogue record for this book is available from the British Library

Additional hard and PDF copies can be obtained from orders@intechopen.com

Importance of Oxidative Stress and Antioxidant System in Health and Disease
Edited by Suna Sabuncuoğlu and Ahmet Yalcinkaya
p. cm.

This title is part of the Biochemistry Book Series, Volume 43
Topic: Chemical Biology
Series Editor: Miroslav Blumenberg
Topic Editors: Şükrü Beydemir and Deniz Ekinci

Print ISBN 978-1-80356-329-9
Online ISBN 978-1-80356-330-5
eBook (PDF) ISBN 978-1-80356-331-2
ISSN 2632-0983

For EU product safety concerns:
IN TECH d.o.o., Prolaz Marije Krucifikse Kozulić 3, 51000 Rijeka, Croatia,
info@intechopen.com or visit our website at intechopen.com.

IntechOpen Book Series
Biochemistry
Volume 43

Aims and Scope of the Series

Biochemistry, the study of chemical transformations occurring within living organisms, impacts all of the life sciences, from molecular crystallography and genetics, to ecology, medicine and population biology. Biochemistry studies macromolecules - proteins, nucleic acids, carbohydrates and lipids –their building blocks, structures, functions and interactions. Much of biochemistry is devoted to enzymes, proteins that catalyze chemical reactions, enzyme structures, mechanisms of action and their roles within cells. Biochemistry also studies small signaling molecules, coenzymes, inhibitors, vitamins and hormones, which play roles in the life process. Biochemical experimentation, besides coopting the methods of classical chemistry, e.g., chromatography, adopted new techniques, e.g., X-ray diffraction, electron microscopy, NMR, radioisotopes, and developed sophisticated microbial genetic tools, e.g., auxotroph mutants and their revertants, fermentation, etc. More recently, biochemistry embraced the 'big data' omics systems. Initial biochemical studies have been exclusively analytic: dissecting, purifying and examining individual components of a biological system; in exemplary words of Efraim Racker, (1913 –1991) "Don't waste clean thinking on dirty enzymes." Today, however, biochemistry is becoming more agglomerative and comprehensive, setting out to integrate and describe fully a particular biological system. The 'big data' metabolomics can define the complement of small molecules, e.g., in a soil or biofilm sample; proteomics can distinguish all the proteins comprising e.g., serum; metagenomics can identify all the genes in a complex environment e.g., the bovine rumen.
This Biochemistry Series will address both the current research on biomolecules, and the emerging trends with great promise.

Meet the Series Editor

Miroslav Blumenberg, Ph.D., was born in Subotica and received his BSc in Belgrade, Yugoslavia. He completed his Ph.D. at MIT in Organic Chemistry; he followed up his Ph.D. with two postdoctoral study periods at Stanford University. Since 1983, he has been a faculty member of the RO Perelman Department of Dermatology, NYU School of Medicine, where he is codirector of a training grant in cutaneous biology. Dr. Blumenberg's research is focused on the epidermis, expression of keratin genes, transcription profiling, keratinocyte differentiation, inflammatory diseases and cancers, and most recently the effects of the microbiome on the skin. He has published more than 100 peer-reviewed research articles and graduated numerous Ph.D. and postdoctoral students.

Meet the Volume Editors

Suna Sabuncuoğlu, professor in the Faculty of Pharmacy, Department of Pharmaceutical Toxicology, Hacettepe University, Turkey, graduated from the same institution, completing a Ph.D. in pharmaceutical toxicology. As a Ph.D. student, she worked at the International Agency for Research on Cancer, Molecular Carcinogenesis Laboratory. She also held a postdoctoral position at the Department of Chemotherapy and Virology, Rega Institute, The Catholic University of Leuven, Belgium. Suna Sabuncuoğlu became a lecturer in 2013 and an associate professor in 2014. She has served on many different boards, commissions and centers within and outside the university. Since 2018, she has held the title of European Registered Toxicologist (ERT).

Ahmet Yalcinkaya, MD, Ph.D., graduated from medical school in 2013 and worked as an ER physician for nearly two years. He began his Ph.D. in the Medical Biochemistry Department of Hacettepe University Faculty of Medicine and defended his thesis in 2020. During his Ph.D., he received a grant for cancer and immunology research at the Medical University of South Carolina, USA, and worked at the Hollings Cancer Center, Department of Biochemistry and Molecular Biology from 2019 to 2020. His current postdoctoral research at the Department of Medical Biochemistry and Microbiology, Uppsala University, Sweden, explores the antibody-mediated origins of immune and infectious diseases.

Contents

Preface

An oxidative imbalance caused by the inability to detoxify the reactive products that are created by the generation of reactive oxygen species (ROS) during cellular metabolism is known as oxidative stress. Increased levels of free radicals produced by biological oxidation lead to cell death, oxidative damage to cellular structures and components, disruption of intracellular protein structure and function, membrane damage from polyunsaturated fatty acid peroxidation with lipids, nucleic acid base modifications, and chromosome changes.

Superoxide and the hydroxyl radical are the two main oxygen free radicals. They are created through chemical reduction processes starting with molecular oxygen. These free radicals can harm cells and cause apoptosis, which can contribute to a number of serious conditions, including diabetes, cancer, myocardial infarction, and stroke. Free radicals and DNA interactions are assumed to be the cause of many malignancies, which subsequently result in mutations that alter the cell cycle and neoplasia. Because free radicals are necessary for life, the body minimizes radically induced damage to protect against excessive production of free radicals by several enzymatic mechanisms. Antioxidants play a crucial role in these defense mechanisms. A careful equilibrium between oxidants and antioxidants allows healthy organisms to resist the damaging effects of ROS. Hence, the steady creation of free radicals in aerobic organisms must be balanced by a constant consumption of antioxidants. Antioxidants, whether they are enzyme- or non-enzyme-based, work to stop free radicals from forming and to find, neutralize, or repair the harm they do. Many natural and exogenous antioxidants are used to protect against oxidative damage and chronic illnesses.

Several antioxidant mechanisms found in both plants and the human body maintain the equilibrium of ROS. The mitochondrial respiratory chain, which naturally produces ROS, plays a role because ROS can, in some circumstances, be both damaging to cells and advantageous to metabolism. In contrast, under pathological or stressful conditions, ROS overwhelm antioxidant systems, resulting in an imbalance that, in turn, leads to oxidative stress and irreversible changes in cell compounds, such as proteins, carbohydrates, and lipids, as well as the ability to interfere with standard cellular-signaling mechanisms. A growing understanding exists that elevated ROS may contribute significantly to the pathogenesis of many chronic diseases, including cancer, atherosclerosis, cardiovascular disease, diabetes, Parkinson's and Alzheimer's disease, liver injury, and immune dysfunction, as well as the process of normal aging.

The potential ability of antioxidants to prevent cancer by neutralizing ROS that might harm DNA has long been a theory. Vitamin E, flavonoids, and polyphenols are a few antioxidants that have been studied in recent years for their potential or claimed benefits against oxidative stress.

Elevated amounts of exogenous antioxidants have been shown to inhibit the kinds of free radicals linked to the emergence of cancer in laboratory and animal studies. In cooperation with the National Cancer Institute, several randomized studies assessing the value of antioxidant supplements for cancer prevention have been carried out. There is no evidence to support their effectiveness in preventing primary cancer. According to a US investigation, there is insufficient solid scientific data to support the benefits of vitamin and mineral supplements in preventing cancer. Antioxidants can help some types of patients who have a real, documented imbalance, but they might not provide any additional benefits for someone who consumes enough nutrients in their diet.

This book provides a guide to the pathophysiology of oxidative stress and the latest therapeutic advances in modulating antioxidants.

Dr. Suna Sabuncuoğlu
Professor,
Faculty of Pharmacy,
Department of Toxicology,
Hacettepe University,
Sıhhiye, Ankara, Turkey

Ahmet Yalcinkaya
Department of Medical Biochemistry and Microbiology,
Uppsala University,
Uppsala, Sweden

Section 1

Oxidative Stress Related Diseases

Chapter 1

The Role of Oxidative Stress in the Onset and Development of Age-Related Macular Degeneration

Emina Čolak, Lepša Žorić, Miloš Mirković, Jana Mirković, Ilija Dragojević, Dijana Mirić, Bojana Kisić and Ljubinka Nikolić

Abstract

Age-related macular degeneration (AMD) is a complex, degenerative and progressive chronic disease that leads to severe visual loss. The prevalence of early AMD accounts for 18% in the population between 65 and 74 years of age and even 30% in subjects older than 74 years. The articles published in the last decade point out to a significant role of oxidative stress in the onset and development of age-related macular degeneration. Generally, reactive oxygen species (ROS) are produced in the eye during light absorption and physiological metabolic processes. The level of oxidative stress is kept under control by the action of antioxidants and reparative enzymes. Excessive synthesis of ROS leads to increased oxidative modification of lipids, proteins and DNA, causing oxidative damage of cytoplasmic and nuclear cell elements and changes of the extracellular matrix. The accumulation of oxidatively modified compounds in drusen deposits will initiate the onset and development of AMD. The objective of this review was to highlight the mechanisms of oxidative stress in order to elucidate their significance and association with the pathogenesis of AMD.

Keywords: age-related macular degeneration, antioxidants, oxidative stress, reactive oxygen species (ROS), retinal pigment epithelium (RPE)

1. Introduction

Age-related macular degeneration (AMD) is a complex, degenerative, progressive, multifactorial disease with multiple genetic and environmental factors contributing to its onset and progression [1].

Age-related macular degeneration (AMD) represents damage of the retinal macula and thus the central visual field and is the leading cause of blindness and visual impairment in people over 60 years of age [2, 3]. Prevalence of an early AMD (the presence of medium-sized drusen or drusen with degeneration or hyperpigmentation of the retinal pigment epithelium (RPE)) is 18% in the elderly population between 65 and 74 years of age and as much as 30% in the population older than 74 years [4].

The initial site of damage, according to most researchers, is the retinal pigment epithelium, although some authors find primary damage in the choriocapillary or extracellular matrix of the sensory retina. Regardless of the location and mechanism of initial damage, there is an opinion that oxidative stress plays an increasingly important role in the genesis of age-related macular degeneration.

The eye is a unique organ, because it is constantly exposed to radiation, atmospheric oxygen, chemicals from the environment, but also to the physical damages [5]. Therefore, oxidative stress is one of the most important mechanisms of the onset of many eye diseases such as cataract, glaucoma, uveitis, retrolental fibroplasia, age-related macular degeneration, as well as various forms of retinopathy [6]. Most free radicals are formed as bioproducts of normal cellular physiology. The most common damage to the eye by free radicals is caused by hydroxyl radical (OH·), superoxide anion radical ($O2^{\bullet-}$) and hydrogen peroxide (H_2O_2). Reactive oxygen species (ROS) cause oxidative damage to cytoplasmic and nuclear elements of the cell and make changes in the extracellular matrix. The degree of oxidative cell damage is limited by the action of various types of antioxidants and reparation of damaged structures. Throughout life, persistent oxidative stress which leads to oxidative damage to macromolecules and to the accumulation of these oxidatively modified compounds is one of the most important factors of tissue ageing. The retina is a typical example of tissue where oxidative changes occur, including loss of retinal cells, accumulation of lipofuscin within the retinal pigment epithelium (RPE), drusen formation, accumulation of degrading products in the Bruch membrane and changes in choroidal capillaries. When these changes become pronounced, they contribute to the formation of macular degeneration. Recent studies have shown that antioxidants and 'scavengers' of free radicals have anti-inflammatory and protective effects on eye tissues, protecting them from the harmful effects of oxidants [7].

The objective of this review was to describe the mechanisms of oxidative stress in order to elucidate their significance and association with the pathogenesis of AMD.

2. The classification of AMD

Based on the Beckman AMD classification system, the disease is classified into early-stage AMD, intermediate-stage AMD and late-stage AMD [8]. Early-stage AMD encompasses the presence of medium-sized drusen (63–125 μm) without any impairment of visual function. Intermediate-stage AMD is defined by the presence of large drusen (>125 μm) or/and abnormalities in the RPE. Late-stage AMD (advanced AMD) is classified into two clinical entities: central geographic atrophy (GA, dry or nonexudative AMD) and neovascular AMD (wet or exudative AMD) [9]. Irreversible loss of vision occurs in geographic atrophy when there is an irreversible loss of RPE and photoreceptor cells, usually in the perifoveal region of the macula. In the neovascular form of AMD, there is an invasion of new choroidal blood vessels (choroidal neovascularization-CNV), followed by retinal detachment and RPE and vision loss [9].

3. Oxidative processes and the onset of AMD

Oxidative processes participate in almost all pathological processes in the eye. The presence of oxidative stress has also been registered in uveitis, diabetic retinopathy, various forms of glaucoma, cataractogenesis and other degenerative processes [10]. As highly reactive intermediates, free radicals can lead to oxidative tissue damage through a number of mechanisms such as peroxidation of unsaturated fatty acids leading to disturbances in the permeability and fluidity of biological membranes, which is accompanied by increased membrane permeability. Oxidation of thiol groups of enzymes leads to a decrease in their activity and even inactivation of enzymes. Fragmentation of fatty acid chains leads to loss of membrane integrity, while disruption of lysosomal membrane continuity leads to release of hydrolytic enzymes and cell damage [11].

The oxidation of polyunsaturated fatty acids in the phospholipids of the cell membranes could damage the cell integrity and function. In addition to negative and destructive effects, this process may have important physiological functions such as: the lipid metabolism regulation, and changes in their physicochemical properties and permeability. Under controlled conditions, ROS enables the control of synthesis of biologically active prostaglandins and leukotrienes, proliferation and initiation of cell death.

Proteins are also targeted by free radicals' action that could change their primary, secondary and even tertiary structure. Oxidative modifications of the primary structure of proteins resulting from the modification or loss of some amino acids or aggregation and fragmentation of proteins, which are reflected in changes in solubility and charge, are described [12]. These processes affect the integrity of the cell and its function and lead to oxidative tissue damage.

Oxidation of nucleic acids leads to changes in DNA structure, gene mutations, synthesis of inadequate genes or lack of synthesis of other genes. As a result of such processes, malignant cell transformations occur. Mitochondrial DNA is particularly sensitive to such transformations.

The degree of biomolecule damage depends on their vulnerability and intensity of oxidative stress. The repair of primarily damaged molecules results in structural changes that remain at the molecular, i.e. at the cellular level. At one point, the damage becomes so great that it exceeds the critical mass. At that moment, the symptoms of illness appear [13].

Eye damages caused by these changes as well as the mechanisms of antioxidant protection show certain specifics, not only in the eye as a special organ but also in its highly differentiated and specialised structures. Oxidative stress in epithelial cells occurs mainly as a consequence of a photodynamic process or as a by-product of oxidative phosphorylation in mitochondria.

The retina is very complex in its structure, and it is one of the highest oxygen-consuming tissues that continuously transforms light into vision, generating reactive oxygen species (ROS), such as the superoxide (O2•−), the hydroxyl radical (•OH), hydrogen peroxide (H_2O_2) and singlet oxygen ($_1O^2$) as normal metabolic by-products [14]. Generally, ROS are produced during oxidative metabolism under physiological conditions and participate in normal cellular metabolism [15]. Retinal photoreceptor membranes are rich in polyunsaturated fatty acids.

Photooxidative retinal damage is in the function of duration of intensity and wavelength of light. Changes that occur in the pigment layer of the retina are considered to be initial in the process of genesis of age-related macular degeneration. During ageing, functions of all senses gradually weaken. Degenerative processes in the eye and especially in the lens are the first signs of ageing that are noticed [10].

4. The synthesis of free radicals during light exposure

It is well established that light exposure has the potential to cause detrimental effects in RPE and retina as well as in many other organs and tissues, such as the skin, cornea, conjunctiva and lens [16].

Large quantities of ROS are produced by exposure to ultraviolet light (λ = 100–400 nm) and to blue light (λ = 400–500 nm) [17]. The photoreceptors in the macula absorb parts of the light spectrum through rhodopsin, a photoreceptor molecule in rods [18].

Roehlecke and Schumann [19] suggested that the synthesis of ROS occurred directly in outer segments of photoreceptors in the reaction catalysed by the enzyme nicotinamide adenine dinucleotide phosphate oxidase (NOX) as well as by the mitochondrial activity of the outer segments after absorption of visible blue light (λ = 405 nm) with an output power of 1 mW/cm^2 [19]. The authors found that the generation of ROS is highly increased in the photoreceptors of retinal explants after 0.5–1 h of blue light absorption, due to increased NOX activity (especially NOX2 and NOX4). Under these conditions (light exposure of 1 mW/cm^2), it is possible to do the following extrapolation to the superoxide anion [20]: 1) One granule of lipofuscin can synthesise 8 x 10^{-19} mol of superoxide anion/min; 2) since 1 mol contains 6.02 x 10^{23} molecules, then 1 granule is capable of producing 4.8 x 10^5 molecules of superoxide/min; and 3) if we take into account that the average cell volume is 2000 μm^3 and if up to 19% of that volume is occupied by lipofuscin granules 1 μm in diameter, then each RPE cell has a synthesis capacity of 3.5 x 10^8 superoxide anions/per cell per minute [21].

This high level of free radical synthesis may explain why RPE cells contain a high concentration of various antioxidants [22]. The spectral dependence of lipofuscin explains the so-called 'blue light hazard' on retina. Light with a wavelength of 550 nm or less can cause 'actin' or photochemical damage, but is too low to cause thermal effects [23]. These photochemical lesions are expressed at the level of RPE, where the action spectrum of 'blue light' is similar to the bandwidth of the absorption spectrum of melanin [24] and lipofuscin [22, 25]. Photoreactivity analysis of blue light in freshly isolated RPE cells shows a high level of oxygen uptake with increasing age of the donor and that this 'photo-uptake' is predominantly related to lipofuscin [26]. These observations suggest a different function of lipofuscin in cells that may explain the association between high levels of lipofuscin and AMD. RPE cells are rich in antioxidants which may be enough to detoxify any reactive oxygen species [27]. Conversely, antioxidants may be insufficient to detoxify all radicals throughout life so that oxidative damage can manifest at some point in life (for example, in old age).

It has also been observed that lipofuscin photosensitivity reactions lead to increased intragranular lipid peroxidation, measured through the accumulation of lipid peroxides and malondialdehyde in pigment granules [26, 28]. Moreover, lipofuscin can perform extracellular lipid peroxidation and enzyme inactivation. Freshly isolated lipofuscin granules incubated with visible light induced up to a 30% increase in lipid peroxidation, compared with the control. Granules incubation with catalase (antioxidant) and lysosomal enzymes (acid phosphatase), in the presence of light, causes as much as 30–50% reduction in enzyme activity. Lipid peroxidation and loss of enzyme activity can be prevented by antioxidants which indicate that lipofuscin photodamage is a product of action of free reactive oxygen species. It is generally accepted that RPE cell dysfunction is an early, crucial moment in the pathogenesis of AMD [29, 30].

RPE cells have a variety of functions from metabolic to supportive, and they are vital for photoreceptors including maintenance of the blood-retinal barrier, participation in the visual cycle (uptake, transport and release of vitamin A and its metabolites) as well as in degradation and uptake of apical phagocited parts of the photoreceptor outer segments [31]. One of the leading factors of RPE cell dysfunction is age-related phagocytic and metabolic insufficiency of postmitotic RPE cells, leading to progressive accumulation of lipofuscin granules which are mainly composed of lipids (~50%) and protein (~44%) of phagosomal, lysosomal and photoreceptor origin (including the retinoid transporter-cellular retinaldehyde binding protein/CRALBP). These substances from the lipofuscin composition can be oxidatively modified either as a result of exposure to UV light or high doses of oxygen in the eye [29, 30].

The well-known cytotoxic constituent of lipofuscin is fluorophore bisretinoid which consists of two retinoid chains derived from the pyridinium ring (A2E) which together with other photoreactive molecules is a powerful photoinducible ROS generator with a strong effect on oxidative damage of lipids, proteins and DNA [31]. N-retinyl-N-retinyldiene-ethanolamine 2-(2,6-dimethyl-8-(2,6,6- trimethyl-1-cyclohexene-1-yl) -1E, 3E, 5E, 7E-octyltetraenyl]-1-(2-hydroxyethyl)-4-[4- methyl-6-(2,6,6-trimethyl] or A2E increases the RPE sensitivity to blue light and exhibits several toxic effects on RPE cells [32, 33]. By the action of light of wavelength, λ = 430 nm, A2E is converted to A2E-epoxide by binding to oxygen. The resulting epoxide can destabilise the membranes of mitochondria and lysosomes [34] and can also inhibit cytochrome oxidase, leading to disruption of electron flow in the respiratory chain [35]. This process, in addition to producing more ROS (reactive oxygen species), reduces the efficiency of energy metabolism. An alternative A2E toxic pathway has been described by Finnemann [36], who in a study with A2E-laden RPE cells demonstrated the presence of destabilised lysosomes, resulting in incomplete digestion of phagocited photoreceptors of the outer segments during 24 h. Since phagocytosis is a circadian regulated process, this will constantly increase the non-degraded phospholipids that are a source of ROS. Mitochondrial destabilisation and incomplete digestion of lipids and proteins caused by lysosome destabilisation lead to increased free radical accumulation. In a closed circle, this mechanism destabilises RPE cells, leading to their loss and this process conditions the initiation of drusen formation [37].

Although lipid peroxidation products are considered to be the main substrates for the genesis of lipofuscin and its cytotoxic constituents, other identified lipofuscin proteins also play a significant role in cytotoxicity [30, 31].

The study of King et al. conducted on the human adult RPE cell line-19 (ARPE-19) revealed that the mitochondrial electron transport chain was an important source of ROS which played a critical role in the death of cells exposed to short-wavelength blue light (425 ± 20 nm) [38].

Except lipofuscin, several other retinal pigments, such as rhodopsin and melanin, were shown to be involved in the oxidative stress process [39]. Grimm et al. reported rhodopsin-mediated blue-light-induced damage in the retina, which occurred after short time exposure to the blue light [40].

In the RPE, lipofuscin is derived primarily from phagocytosis of shed photoreceptor outer segments and is considered a heterogeneous waste material that accumulates with age in active postmitotic cells, such as those of the RPE [41]. The RPE cells are able to phagocyte the photoreceptors of outer segments (POSs) that contain a high amounts of unsaturated fatty acids [42]. During phagocytosis, a high quantity of oxygen is consumed and a significant production of ROS occurs (generated by NOX or peroxidase) via the oxidation of fatty acids in the POSs [43]. Mitter et al. in their study have shown that autophagy plays a significant role in protection of the RPE from oxidative stress [44]. Recent evidence showed that dysfunctional autophagy/mitophagy in the RPE may lead to mitochondrial disintegration by affecting the mitochondrial fission/fusion ratio, resulting in excessive amounts of ROS [45].

5. The synthesis of oxidatively modified compounds in the eye, by the action of free radicals

Excessive synthesis of free radicals in the eye is associated with the production of oxidatively modified compounds and cytotoxic damage of ocular structures. The cell types with relatively high levels of polyunsaturated fatty acids (PUFAs), such as retinal cells, are highly sensitive to lipid peroxidation. Polyunsaturated fatty acids in phospholipids and glycolipids are the basic substrate of oxidative damage to lipids caused by free radicals.

Lipid peroxidation, a complex process involving the interaction of oxygen-derived free radicals with polyunsaturated fatty acids, finally results in a variety of primary compounds: highly reactive compounds (alkyl radicals, conjugated dienes, peroxy and alkoxy/oxyl radicals and lipid hydroperoxide). During further decomposition of primary compounds, a series of secondary products are produced such as: short-chain evaporable hydrocarbons, aldehydes and end products of lipid peroxidation (i.e. isoprostanes, MDA, 4-hydroxy-2,3, trans nonenal and 4,5-dihydroxydecenal) [46, 47].

MDA is a secondary product of peroxidation of unsaturated fatty acids, (particularly arachidonic acid) and is a physiological ketoaldehyde [48]. In a higher concentration, it reacts with free amino groups of proteins (especially with lysine cysteine or histidine residue). Such modified protein structures have immunogenic features. Some studies have shown that an increased titre of these autoantibodies directly correlates with the extent of oxidative damage and may predict the progression of some diseases. It was suggested that reduced ability to protein proteolysis after their oxidative modification with MDA and 4-HNE represents one of the main factors of lipofuscin synthesis during the development of AMD [49].

Lipid peroxidation highly reactive end products, such as 4-hydroxylnonenal (4-HNE), malondialdehyde (MDA), oxidised nucleotides and carboxyethyl pyrrole (CEP), have been demonstrated to be associated with drusen formation and RPE atrophic modifications in both human and animal eye [50].

Recently, Kim et al. [51] found that the injection of hydroperoxy-octadecadienoic acid (HpODE), (a peroxidized lipid) into the subretinal space of a murine AMD model, could initiate an early increase in the expression of markers of oxidative stress and lipid peroxidation, especially high levels of 4-HNE and MDA [51]. Zor et al. [52] documented a significantly increased MDA values (~15%) in patients with neovascular AMD compared with the controls [53].

F2-isoprostane (F2-IsoPs) is another marker of lipid peroxidation which is considered to be an important 'in vivo' marker of oxidative damage in AMD [54, 55]. Sabanayagam et al. [56] demonstrated that the presence of F2-IsoPs in urine was positively associated with AMD.

Oxidative DNA damage of both nuclear and mitochondrial genomes can result in strand breaks, base modifications and DNA-protein cross linkages which are all strongly implicated in ageing and age-related diseases [57, 58]. Over 20 base modifications related to ROS attack of DNA are identified, with the following oxidative DNA damage products: 8-oxo-7,8-dihydroadenine, 8-oxo-7,8-dihydroguanine, 8-oxo-7,8-dihydro-2′-deoxyguanosine (8-oxodG) and 5,6-dihydroxy-5,6-dihydrothymine as well as the ring-opened lesions of 4,6-diamino-5-formamido-pyrimidine and 2,6-diamino-4-hydroxy-5-formamido-pyrimidine [59]. The 8-oxodG, which is formed through the oxidation of guanine at the C8 position in the guanine base, serves as a reliable biomarker of oxidative stress and oxidative modification of DNA, and it is associated with ageing and ageing-related diseases [58].

Age-related increases in lipofuscin, 8-oxoguanine, CEP, 4-HNE and MDA expression have been observed in the ageing retina [60–62] which have been reported to cause inflammatory responses and AMD features [63].

6. The pathophisiology of AMD-Drusen formation

Age-related macular degeneration is characterised by degenerative changes involving the outer portion of the retina, RPE, Bruch's membrane and choriocapillaris. Drusen are considered as a hallmark of AMD and as an amorphous deposit that accumulates extracellularly in the zone between the RPE and the inner collagen zone of the Bruch's membrane [64]. Clinically, they are divided into two main phenotypes: 'soft and hard', depending on their relative size and shape. A few smaller hard drusen (<65 μm) can be found in at least 95% of the elderly population, but do not represent AMD. Only the presence of larger drusen (>125 μm), especially soft drusen (> 125–250 μm) in the macula, is considered as a major risk factor for the development of advanced forms of AMD, i.e. exudative-neovascular forms, especially if they are combined with pigmentation disorders [65].

In the later stage of AMD, neovascularization, exudative changes or disciform scars can occur. In the atrophic form of AMD, there is a loss of pigment epithelium or 'attenuation' of the choriocapillaris but without neovascularization [66]. Early pathological changes include basal deposits in the Bruch's membrane which occur exclusively in pathological samples and have two types: a) basal laminar deposits consisting of basement membrane proteins and long collagen filaments located between RPE and basement membrane and b) basal linear deposits that are more specific for early AMD changes and consist mainly of membrane material located in the Bruch membrane, externally from the RPE basement membrane. The combination of these deposits with secondary changes in RPE results in the formation of drusen [67].

Many different molecules have been identified in drusen, including glycoconjugates and other compounds also found in atherosclerotic plaque (hence the link between atherosclerosis and AMD formation by some authors), including vitronectin, apoprotein B and E, α-crystalline, HtrA1 and lipids [68–70]. Macrophages found in drusen regression suggest a possible hypothesis that macrophages are involved in the process of degradation of deposits within the Bruch membrane [71]. Activated microglias have also been found in AMD degenerative lesions [72]. Discrete nodules or hard drusen deposits consisting of hyaline-like material were found between the RPE and the Bruch membrane. Soft drusen are usually large and occur with detachment of RPE cells and diffuse changes of the Bruch membrane. They can occur in deeper damages of the RPE and choroids and lead to choroidal neovascularization or cell death in the RPE as well as geographical atrophy. Autofluorescent pigments, such as lipofuscin, which are accumulated in RPE cells, reach a size that often leads to decreased cellular function, retinal ageing and degeneration, mostly in the form of geographic atrophy [73].

Lipofuscin in RPE is the most common cause of fundus autofluorescence. These are spherical particles of micrometre size with characteristic yellow fluorescence when exposed to blue light. The main component of lipofuscin is N-retinylidine-N-retinyletanol-amine (A2E), a quaternary amine and retinoid bioproduct of visual cycle [74]. Lipofuscin synthesis is a pathogenic reaction in which the resulting A2E interferes with the function of RPE cells and leads to their apoptosis. Choroidal neovascularization can occur in the macular, peripapillary and peripheral regions. Early choroidal neovascularization occurs below the RPE cells to later break through the RPE layer and develop an exudative, haemorrhagic or disciform form of AMD. In the neovascular form of AMD, lipid accumulation occurs below the RPE or neuroretin. In the haemorrhagic form of AMD, blood penetrates through the RPE into the subretinal space and sometimes through the retina to the vitreous. In the disciform form of AMD, fibrous tissue with neovascularization and changes in RPE cells proliferates and may partially or totally replace neuroretin [75]. Additional pathological lesions include serous exudation, haemorrhage, gliosis and calcification. Macrophages have been proven both morphologically and functionally in the neovascular form of AMD [76]. Activated macrophages and microglias can secrete chemokines and cytokines, causing further cell damage, degradation of the Bruch's membrane and angiogenesis [77].

7. Oxidative stress and choroidal vascular changes in AMD

Among AMD cases, approximately 10–15% have neovascular AMD characterised by abnormal vascular morphology and growth [8]. Vascular endothelial growth factor (VEGF) upregulation plays a crucial role in the development of neovascular AMD. Yi and assoc. [78] documented an increased VEGF expression in a study using laser to induce choroidal neovascularization (CNV) in rats. This author suggests that the macrophages could be probably the most important source of VEGF in the early phase of AMD [78]. VEGF expression in subfoveal fibrovascular membranes was concentrated in cells resembling fibroblasts, implicating a significant role of fibroblasts in the progression of CNV [79]. The results showed that even temporary overexpression of VEGF in RPE cells was sufficient to induce CNV in the rat eye [80]. Wang et al. reported that IQ protein motif-containing GTPase activating protein 1 (IQGAP1), scaffold protein with a Rac1-binding domain, regulated VEGF activation by binding to Rac1GTP in choroidal endothelial cells, activating their migration [81]. IQ motif-containing GTPase-activating protein 1 (IQGAP1) is a ubiquitously expressed scaffold protein that is involved in multiple cellular functions such as cell survival and trafficking [82].

The vascular endothelial dysfunction is considered as a crucial event in development and progression of choroidal vascular dysfunction [82]. Nitric oxide and nitric oxide synthase enzymes have been shown to be involved in the upregulation of VEGF. Nitric oxide synthases (NOSs) are a family of enzymes that catalyse the conversion of L-arginine into nitric oxide (NO). They are classified into three isoforms: endothelial NOS (eNOS), neuronal NOS (nNOS) and inducible NOS (iNOS) [83]. The eNOS maintains the physiological function of the vascular endothelium [84]. It was demonstrated that eNOS mediated endothelium-dependent vasodilation in retinal arterioles and ophthalmic arteries [85]. NO is considered not only a mediator of vasodilation, but also a regulator of various vascular functions. For example, physiologically, NO can dilate a blood vessel, inducing relaxation of vascular smooth muscle cells

(VSMCs), inhibiting cell proliferation and regulating angiogenesis and vascular permeability [86]. Bhutto et al. [87] reported that eNOS and nNOS expression was significantly decreased in the eyes of AMD patients. This author suggested that the decreased expression of eNOS and nNOS might reduce the NO production that could induce hemodynamic changes in CNV [87].

It was documented that excessive amounts of NO can have detrimental effects on cells and tissues, implicated that the production of NO is not always beneficial. In that case, NO can be an important stimulator of CNV. Ando et al. suggested that blockade of nNOS and iNOS could reduce CNV formation [88]. Excessive amounts of nitric oxide can react with the superoxide anion to form peroxynitrite, a very toxic and reactive radical which compromises vascular endothelial function [89].

There is some evidence that ROS and vascular dysfunction may together contribute to the pathology of neovascular AMD. It was demonstrated that NOX was the connection between VEGF and ROS in human choroidal endothelial cells [90]. The family of NOX consists of seven isoforms such as: NOX1, NOX2, NOX3, NOX4, NOX5, dual oxidase (Duox) 1 and Duox2) which are differentially expressed in tissues and cells [91]. NOX1, NOX2 and NOX4 are expressed in choroidal vascular endothelial cells [92]. ROS generated by NOX function as signalling molecule promoting endothelial cell proliferation, migration and tube formation [92]. Some studies documented that ROS generated from NOX2 could activate the transcription factors NF-κB and activator protein 1 (AP-1) and increase the expression of intracellular adhesion molecule (ICAM)-1 and VEGF leading to vascular hyperpermeability and retinal neovascularization [93]. Moreover, NOX4-derived ROS generation is essential for the expression of hypoxia-inducible factor 1-alpha (HIF-1α) which was linked to cell proliferation and migration of vascular smooth muscle cells [94].

8. Other mechanisms involved in the development of neovascular AMD

One of the risk factors for AMD may be increased collagen synthesis in the choriocapillaris which is then incorporated into the Bruch's membrane, creating thickenings that precede the appearance of linear deposits [95]. Chromatographic analysis of the drusen showed that they contained more than a hundred different proteins originated from retinal pigment epithelial cells, neuronal retinas and choriocapillaris. However, their composition differs depending on the existence i.e. absence of AMD. It is thought that certain ingredients can promote angiogenesis. The integrity of the RPE cellular structures in a culture that is chronically exposed to oxidative stress is impaired by the action of hydrogen peroxide due to the interruption of intercellular compounds. This is one of the possible mechanisms of breaking the blood-brain barrier in the pathogenesis of AMD [96].

Programmed cell death (apoptosis) is an essential protective mechanism of the organism against the accumulation and spread of damaged or unnecessary cells. An increased degree of apoptosis is observed in most ageing cell populations. A similar thing happens in RPE cells. There is an opinion that mitochondria play a key role in the regulation of apoptosis. Reactive oxygen metabolites that are formed in RPE cells exposed to the blue part of the spectrum originate in mitochondrial processes [97]. Oxidative stress can reduce the sensitivity of senescent cells to apoptosis through defective oxidative phosphorylation. The process of drusen formation is very similar to apoptotic process in the retina and predisposes the development of neovascularization during the progression of AMD [98].

In postmitotic tissues, during ageing, the oxidatively modified and damaged mitochondrial DNA are accumulated in mitochondria. It is believed that their genetic material is the main substrate of oxidative damage in the retinal pigment epithelium. With inefficient damage repair, redox potential of mitochondria in the human RPE retinal cells is compromised over time in photoreceptors as well [99].

It was suggested that other types of regulated cell death (e.g. pyroptosis, necroptosis and autophagy) may contribute to development of AMD [100]. Ferroptosis is a newly discovered, iron-dependent, regulated cell death pathway that is initiated by lipid peroxidation. It is implicated in neurodegeneration, ischemia–reperfusion injury and myocardial infarction [101]. It is characterised by iron-dependent accumulation [102, 103]. In contrast to apoptosis, ferroptosis is a pro-inflammatory condition that arises due to the release of intracellular content after the rupture of plasma membrane [104]. Under normal conditions, ferroptosis is a mechanism that protects cellular integrity, but leads to cell death when cellular integrity is compromised, while apoptosis represents a suicide mechanism that eliminates certain types of cells from the whole organism at specific time points [105].

It was documented that angiotensin II (Ang II) was implicated in the pathology of AMD. It was shown that Ang II can mediate various pathological processes in ocular blood vessels such as proliferation and migration of smooth muscle cells and pericytes, increase of VEGF expression and potentiation of VEGF-dependent angiogenic activity [106, 107]. Receptors for AngII have been identified in retinal and optic nerve blood vessels. Some studies have shown that blocking the renin-angiotensin system may delay the breakdown of the blood-retinal barrier and prevent retinal neovascularization and the development of AMD [108].

9. Conclusion

In this review, we tried to highlight the pathways of oxidative stress and their implication in the pathogenesis of AMD. Considering the unique structure and function of the retina in the eye, as well as the environment in which it is located, it indicates a significant synthesis of free radicals during normal physiological processes as well as during light absorption. The presence of free fatty acids and their exposure to free radicals make lipid peroxidation processes a daily occurrence in the eye. This was confirmed by many studies that found high concentrations of MDA, 4-HNE and other lipid peroxidation products in the eyes (and blood) of AMD patients. Oxidative damage of mitochondria and nuclear DNA was also observed in AMD patients, as well as increased products of oxidative damage of proteins. An impairment of autophagy and other types of cell death such as pyroptosis, necroptosis and ferroptosis were also described in AMD patients. The upregulation of VEGF and isoforms of NOX with impairment of NO synthesis have significant implications in the development of new blood vessels and the onset of choroidal neovascularization (CNV) in the pathogenesis of advanced-wet AMD. In view of all the above, further research is certainly needed in order to find adequate methods for disease prevention as well as adequate drugs for the treatment of various forms of AMD.

Author details

Emina Čolak[1*], Lepša Žorić[2,3], Miloš Mirković[2,3], Jana Mirković[2,3], Ilija Dragojević[4], Dijana Mirić[4], Bojana Kisić[4] and Ljubinka Nikolić[5]

1 University Clinical Centre of Serbia, Institute of Medical Biochemistry, Belgrade, Serbia

2 Faculty of Medicine, Ophthalmology Department, University of Priština, Settlement Kososvska Mitrovica, Serbia

3 University Clinical Centre of Serbia, Clinic for Eye Diseases, Belgrade, Serbia

4 Faculty of Medicine, Biochemistry Department, University of Priština, Settlement Kososvska Mitrovica, Serbia

5 Department for Haematology and Transfusion, University Clinical Centre of Serbia, Clinic for Gynaecology and Obstetrics, Belgrade, Serbia

*Address all correspondence to: emina.colak.bg@gmail.com

References

[1] Čolak E, Majkić-Singh N, Žorić L, Radosavljević A, Kosanović-Jaković N. The role of CRP and inflammation in the pathogenesis of age-related macular degeneration. Biochemia Medica (Zagreb). 2012;**22**:39-48

[2] Klein R, Klein BE, Jensen SC, Meuer SM. The five-year incidence and progression of age-related maculopathy: The beaver dam eye study. Ophthalmology. 1997;**104**:7-21

[3] Georgijevic A, Tomic Z. Primena fototerapije u lecenju subfovealne horoidalne neovaskularizacije. Srpski Arhiv za Celokupno Lekarstvo. 2007;**135**(11-12):629-634

[4] Čolak E, Ignjatović S, Radosavljević A, Žorić L. The association of enzymatic and non-enzymatic antioxidant defense parameters with inflammatory markers in patients with exudative form of age-related macular degeneration. Journal of Clinical Biochemistry and Nutrition. 2017;**60**(2):100-107. DOI: 10.3164/jcbn.16-30

[5] Ozawa Y. Oxidative stress in the light-exposed retina and its implication in age-related degeneration. Redox Biology. 2020;**37**:2213-2217

[6] Ohia SE, Opere CA, LeDay AM. Pharmacological consequences of oxidative stress in ocular tissues. Mutation Research. 2005;**579**:22-36

[7] Bosch-Morell F, Roma J, Marin N, Romero B, Rodriguez-Galietero A, Johnsen-Soriano S, et al. Role of oxygen andnitrogen species in experimental uveitis: Anti-inflammatory activity ohthe synthetic antioxidant ebselen. Free Radical Biology & Medicine. 2002;**33**:669-667

[8] Ruan Y, Jiang S, Gericke A. Age-related macular degeneration: Role of oxidative stress and blood vessels. International Journal of Molecular Sciences 2021;**22**:1296. DOI: org: 10.3390/ijms22031296

[9] Ferris FL, Wilkinson CP. Clinical classification of age-related macular degeneration. Ophthalmology. 2013;**120**:844-851

[10] Žoric L, Jovanovic P. Oksidativni stres u patogenezi ocnih bolesti (in Serbian). Niš: Sven; 2006

[11] Djordjevic V, Pavlovic D, Kocic G. Biohemija slobodnih radikala. Medicinski fakultet Nis. 2000. p. 320

[12] Čolak E. New markers of oxidative damage to macro molecules. Journal of Medicine Biochemistry. 2008;**27**(1): 1-16

[13] Thomas JP, Maivrino M, Ursini F, Girrotti AW. Protective action of phospholipid hydroperoxide glutathione peroxidase against membrane damaging lipid peroxidation. The Journal of Biological Chemistry. 1990;**265**:454-461

[14] Yu DY, Cringle SJ. Retinal degeneration and local oxygen metabolism. Experimental Eye Research. 2005;**80**:745-751

[15] Ray PD, Huang BW. Reactive oxygen species (ros) homeostasis and redox regulation in cellular signaling. Cellular Signalling. 2012;**24**:981-990

[16] Marek V, Mélik-Parsadaniantz S. Blue light phototoxicity toward human corneal and conjunctival epithelial cells in basal andhyperosmolar conditions.

Free Radical Biology & Medicine. 2018;**126**:27-40

[17] Nakashima Y, Ohta S. Blue light-induced oxidative stress in live skin. Free Radical Biology & Medicine. 2017;**108**:300-310

[18] Funk RH, Schumann U. Blue light induced retinal oxidative stress: Implications for macular degeneration. World Journal of Ophthalmology. 2014;**4**:29-34

[19] Roehlecke C, Schumann U. Stress reaction in outer segments of photoreceptors after blue light irradiation. PLoS One. 2013;**8**:e71570

[20] Boulton M, Dontsov A, Jarvis-Evans J, Ostrovsky M, Svistunenko D. Lipofuscin is a photo inducible free radical generator. Journal of Photochemistry and Photobiology. B. 1993;**19**:201-204

[21] Feeney-Burns L, Hilderbrand ES, Eldridge S. Aging human RPE: Morphometric analysis of macular, equatorial, and peripheral cells. Investigative Ophthalmology & Visual Science. 1984;**25**:195-200

[22] Boulton ME. Ageing of the retinal pigment epithelium. Progress in Retinal and Eye Research. 1991;**11**:125-151

[23] Ham WT Jr, Ruffolo JJ Jr, Mueller HA, Guerry D, 3d. The nature of retinal radiation damage: Dependence on wavelength, power level and exposure time. Vision Research. 1980;**20**:1105-1111

[24] Marshall J. Radiation and the ageing eye. Ophthalmic & Physiological Optics. 1985;**5**:241-263

[25] Eldred GE, Katz ML. Fluorophores of the human retinal pigment epithelium: Separation and spectral characterization. Experimental Eye Research. 1988;**47**:71-86

[26] Rozanowska M, Jarvis-Evans J, Korytowski W, Boulton ME, Burke JM, Sarna T. Blue light-induced reactivity of retinal age pigment. In vitro generation of oxygen-reactive species. The Journal of Biological Chemistry. 1995;**270**:18825-18830

[27] Winkler BS, Solomon FJ, Orselli SM. Effects of oxidized glutathione on ATPase activities in rat retina. Investigative Ophthalmology & Visual Science. 1991;**32**(10):2840-2842

[28] Wassell J, Davies S, Bardsley W, Boulton M. The photo reactivity of the retinal age pigment lipofuscin. The Journal of Biological Chemistry. 1999;**274**(34):23828-23832

[29] Sparrow JR, Boulton M. RPE, lipofuscin and its role in retinal pathobiology. Experimental Eye Research. 2005;**80**:595-606

[30] Warburton S, Southwick K, Hardman RM, Secrest AM, Grow RK, Xin H, et al. Examining the proteins of functional retinal lipofuscin using analysis as a guide for understanding its origin. Molecular Vision. 2005;**11**:1122-1134

[31] Strauss O. The retinal pigment epithelium in visual function. Physiological Reviews. 2005;**85**:845-881

[32] Sparrow JR, Fishkin N, Zhou J, Cai B, Jang YP, Krane S, et al. A2E, a byproduct of the visual cycle. Vision Research. 2003;**43**:2983-2990

[33] Sparrow JR, Vollmer-Snarr HR, Zhou J, Jang YP, Jockusch S, Itagaki Y, et al. A2E-epoxides damage DNA in retinal pigment epithelial cells. Vitamin E and other antioxidants

inhibit A2E-epoxide formation. The Journal of Biological Chemistry. 2003;**278**:18207-18213

[34] Bergmann M, Schutt F, Holz FG, Kopitz J. Inhibition of the ATP drivenproton pump in RPE lysosomes by the major lipofuscin fluorophore A2-E may contribute to the pathogenesis of age-related macular degeneration. The FASEB Journal. 2004;**18**:562-564

[35] Shaban H, Gazzotti P, Richter C. Cytochrome c oxidase inhibition by Nretinyl-N-retinylidene ethanolamine, a compound suspected to causeage-related macular degeneration. Archives of Biochemistry and Biophysics. 2001;**394**:111-116

[36] Finnemann SC, Leung LW, Rodriguez-Boulan E. The lipofuscin component A2E selectively inhibits phagolysosomal degradation of photoreceptor phospholipid by the retinal pigment epithelium. Proceedings of the National Academy of Sciences USA. 2002;**99**:3842-3847

[37] Ambati J, Ambati BK, Yoo SH, Ianchulev S, Adamis AP. Age-related macular degeneration: Etiology, pathogenesis, and therapeutic strategies. Survey of Ophthalmology. 2003;**48**:257-293

[38] King A, Gottlieb E. Mitochondria-derived reactive oxygen species mediate blue light-induced death of retinal pigment epithelialcells. Photochemistry and Photobiology. 2004;**79**:470-475

[39] Sparrow JR, Nakanishi K. The lipofuscin fluorophore a2e mediates blue light–induced damage to retinal pigmented epithelialcells. Investigative Ophthalmology and Visual Science. 2000;**41**:1981-1989

[40] Grimm C, Wenzel A. Rhodopsin-mediated blue-light damage to the rat retina: Effect of photoreversal of bleaching. Investigative Ophthalmology and Visual Science. 2001;**42**:497-505

[41] Sparrow JR, Boulton M. Rpelipofuscin and its role in retinal pathobiology. Experimental Eye Research. 2005;**80**:595-606

[42] Naso F, Intartaglia D. Light-responsive microrna mir-211 targets ezrin to modulate lysosomal biogenesis and retinal cell clearance. The EMBO Journal. 2020;**39**:e102468

[43] Zhang ZY, Bao XL. Autophagy in age-related macular degeneration: A regulatory mechanism of oxidative stress. Oxidative Medicine and Cellular Longevity. 2020;**2896036**:13

[44] Mitter SK, Song C. Dysregulated autophagy in the rpe is associated with increased susceptibility to oxidative stress and amd. Autophagy. 2014;**10**:1989-2005

[45] Youle RJ, van der Bliek AM. Mitochondrial fission, fusion, and stress. Science. 2012;**337**:1062-1065

[46] Kopitz J, Holz FG, Kaemmerer E, Schutt F. Lipids and lipid peroxidation products in the pathogenesis of age-related macular degeneration. Biochimie. 2004;**86**(11):825-831

[47] Čolak E, Ignjatović S, Radosavljević A, Žorić L. The association of enzymatic and non–enzymaticantioxidant defense parameters withinflammatory markers in patients with exudativeform of age–related macular degeneration. Journal of Clinical Biochemistry and Nutrition. 2017;**60**(2):100-107

[48] Dalle-Donne I, Rossi R, Colombo R, Giustarini D, Milzani A. Biomarkers of oxidative damage in human disease. Clinical Chemistry. 2006;**52**:601-623

[49] Holz FG, Kaemmerer E, Bindewald A, Kopitz J. Effets of lipidperoxydation-related protein modifications on RPE lysosomal functions, ROS phagocytosis and their impact on lipofuscino genesis. Investigative Ophthalmology & Visual Science. 2004;**45**:3385

[50] Toma C, De Cilla S, Palumbo A, Praveen Garhwal D, Grossini E. Oxidative and Nitrosative stress in age-related macular degeneration: A review of their role in different stages of disease. Antioxidants (Basel). 2021;**10**(5):653. DOI: 10.3390/antiox10050653

[51] Kim S, Kambhampati S, Bhutto IA, McLeod DS, Lutty GA, Kannan RM. Evolution of oxidative stress, inflammation and neovascularization in the choroid and retina in a subretinal lipid induced age-related macular degeneration model. Experimental Eye Research. 2021;**203**:108391. DOI: 10.1016/j.exer.2020.108391

[52] Zor RK, Erşan S, Küçük E, Yıldırım G, Sarı İ. Serum malondialdehyde, monocyte chemoattractant protein-1, and vitamin C levels in wet type age-related macular degeneration patients. Therapeutic Advances in Ophthalmology. 2020;**12**:2515841420951682. DOI: 10.1177/2515841420951682

[53] Ye F, Kaneko H, Hayashi Y, Takayama K, Hwang SJ, Nishizawa Y, et al. Malondialdehyde induces autophagy dysfunction and VEGF secretion in the retinal pigment epithelium in age-related macular degeneration. Free Radical Biology & Medicine. 2016;**94**:121-134. DOI: 10.1016/j.freeradbiomed.2016.02.027

[54] Dammak A, Huete-Toral F, Carpena-Torres C, Martin-Gil A, Pastrana C, Carracedo G. From oxidative stress to inflammation in the posterior ocular diseases: Diagnosis and treatment. Pharmaceutics. 2021;**13**(9):1376. DOI: 10.3390/pharmaceutics13091376

[55] Dismuke WM, Liang J, Overby DR, Stamer WD. Concentration-related effects of nitric oxide and endothelin-1 on human trabecular meshwork cell contractility. Experimental Eye Research. 2014;**120**:28-35. DOI: 10.1016/j.exer.2013.12.012

[56] Sabanayagam C, Lye WK, Januszewski A, Abdul RBBM, GCM C, Kumari N, et al. Urinary Isoprostane levels and age-related macular degeneration. Investigative Ophthalmology and Visual Science. 2017;**58**:2538-2543. DOI: 10.1167/iovs.16-21263

[57] Jarrett SG, Boulton ME. Consequences of oxidative stress in age-related macular degeneration. Molecular Aspects of Medicine. 2012;**33**(4):399-417. DOI: 10.1016/j.mam.2012.03.009

[58] Bohr VA, Ottersen OP, Tonjum T. Genome instability and DNA repair in brain, ageing and neurological disease. Neuroscience. 2007;**145**(4):1183-1186

[59] Breen AP, Murphy JA. Reactions of oxyl radicals with DNA. Free Radical Biology & Medicine. 1995;**18**(6):1033-1077

[60] Zhao T, Guo X, Sun Y. Iron accumulation and lipid peroxidation in the aging retina: Implication of Ferroptosis in age-related macular degeneration. Aging and Disease. 2021;**12**(2):529-551

[61] Nag TC, Kumar P. Wadhwa S age related distribution of 4-hydroxy 2-nonenal immunoreactivity in human

retina. Experimental Eye Research. 2017;**165**:125-135

[62] Hollyfield JG, Bonilha VL, Rayborn ME, Yang X, Shadrach KG, Lu L, et al. Oxidative damage-induced inflammation initiates age-related macular degeneration. Nature Medicine. 2008;**14**:194-198

[63] Nag TC, Wadhwa S, Alladi PA, Sanyal T. Localization of 4-hydroxy 2-nonenal immunoreactivity in aging human retinal Muller cells. Annals of Anatomy. 2011;**1**(93):205-210

[64] Waldstein SM, Vogl WD, Bogunovic H, Sadeghipour A, Riedl S, Schmidt-Erfurth U. Characterization of Drusen and Hyperreflective foci as biomarkers for disease progression in age-related macular degeneration using artificial intelligence in optical coherence tomography. JAMA Ophthalmology. 2020;**138**(7):740-747. DOI: 10.1001/jamaophthalmol.2020.1376

[65] Mano F, Sprehe N, Olsen TW. Association of Drusen Phenotype in age-related macular degeneration from human eye-Bank eyes to disease stage and cause of death. Ophthalmol Retina. 2021;**5**(8):743-749. DOI: 10.1016/j.oret.2020.11.011

[66] Chen L, Messinger JD, Ferrara D, Freund KB, Curcio CA. Stages of Drusen-associated atrophy in age-related macular degeneration visible via histologically validated fundus autofluorescence. Ophthalmol Retina. 2021;**5**(8):730-742. DOI: 10.1016/j.oret.2020.11.006

[67] Lamin A, Dubis AM, Sivaprasad S. Changes in maculardrusen parameters preceding the development of neovascular age-relatedmacular degeneration. Eye (London, England). 2019;**33**(6):910-916. DOI: 10.1038/s41433-019-0338-x

[68] Lin MK, Yang J, Hsu CW, Gore A, Bassuk AG, Brown LM, et al. Tsang SH.HTRA1, an age-relatedmacular degeneration protease, processes extracellular matrix proteins EFEMP1 and TSP1. Aging Cell. 2018;**17**(4):e12710. DOI: 10.1111/acel.12710

[69] Hu ML, Quinn J, Xue K. Interactions between apolipoprotein E metabolism and retinal inflammation in age-related macular degeneration. Life (Basel). 2021;**11**(7):635. DOI: 10.3390/life11070635

[70] Handa JT, Cano M, Wang L, Datta S, Liu T. Lipids, oxidized lipids, oxidation-specific epitopes, and age-related macular degeneration. Biochimica et Biophysica Acta - Molecular and Cell Biology of Lipids. 2017;**1862**(4):430-440. DOI: 10.1016/j.bbalip.2016.07.013

[71] Allingham MJ, Loksztejn A, Cousins SW, Mettu PS. Immunological aspects of age-related macular degeneration. Advances in Experimental Medicine and Biology. 2021;**1256**:143-189. DOI: 10.1007/978-3-030-66014-7_6

[72] Singh SR, Lupidi M, Mishra SB, Paez-Escamilla M, Querques G, Chhablani J. Unique optical coherence tomographic features in age-relatedmacular degeneration. Survey of Ophthalmology. 2020;**65**(4):451-457. DOI: 10.1016/j.survophthal.2020.01.001

[73] Tong Y, Ach T, Curcio CA, Smith RT. Hyperspectral autofluorescence characterization of drusen and sub-RPE deposits in age-relatedmacular degeneration. Annals of Eye Science. 2021;**6**:4. DOI: 10.21037/aes-20-12

[74] Wu Y, Yanase E, Feng X, Siegel MM, Sparrow JR. Structural characterization of bisretinoid A2E photocleavage products and implications for age-related

macular degeneration. Proceedings of the National Academy of Sciences of the United States of America. 2010;**107**(16):7275-7280. DOI: 10.1073/pnas.0913112107

[75] Thomas CJ, Mirza RG. Gill MKAge-related macular degeneration. The Medical Clinics of North America. 2021;**105**(3):473-491. DOI: 10.1016/j.mcna.2021.01.003

[76] Kauppinen A, Paterno JJ, Blasiak J, Salminen A, Kaarniranta K. Inflammation and its role in age-related macular degeneration. Cellular and Molecular Life Sciences. 2016;**73**(9):1765-1786. DOI: 10.1007/s00018-016-2147-8

[77] Kato Y, Oguchi Y, Omori T, Shintake H, Tomita R, Kasai A, et al. Complement activation products and cytokines in Pachychoroid Neovasculopathy and Neovascular age-related macular degeneration. Investigative Ophthalmology & Visual Science. 2020;**61**(13):39. DOI: 10.1167/iovs.61.13.39

[78] Yi X, Ogata N. Vascular endothelial growth factor expression in choroidal neovascularization in rats. Graefe's Archive for Clinical and Experimental Ophthalmology. 1997;**235**:313-319

[79] Kvanta A, Algvere PV. Subfoveal fibrovascular membranes in age-related macular degeneration express vascular endothelial growth factor. Investigative Ophthalmology and Visual Science. 1996;**37**:1929-1934

[80] Spilsbury K, Garrett KL. Overexpression of vascular endothelial growth factor (VEGF) in the retinal pigment epithelium leads to the development of choroidal neovascularization. The American Journal of Pathology. 2000;**157**:135-144

[81] Wang H, Ramshekar A. IQGAP1 causes choroidal neovascularization by sustaining VEGFR2-mediated Rac1 activation. Angiogenesis. 2020;**23**:685-698

[82] Yeo NJY, Chan EJJ. Choroidal neovascularization: Mechanisms of endothelial dysfunction. Frontiers in Pharmacology. 2019;**10**:1363

[83] Forstermann U, Munzel TJC. Endothelial nitric oxide synthase in vascular disease: From marvel to menace. Circulation. 2006;**113**:1708-1714

[84] Heiss C, Rodriguez-Mateos A. Central role of enos in the maintenance of endothelial homeostasis. Antioxidants & Redox Signaling. 2015;**22**:1230-1242

[85] Gericke A, Wolff I. Retinal arteriole reactivity in mice lacking the endothelial nitric oxide synthase (enos) gene. Experimental Eye Research. 2019;**181**:150-156

[86] Albrecht EWJA, Stegeman CA. Protective role of endothelial nitric oxide synthase. The Journal of Pathology. 2003;**199**:8-17

[87] Bhutto I, Baba T. Low nitric oxide synthases (NOS) in eyes with age-related macular degeneration (AMD). Experimental Eye Research. 2009;**90**:155-167

[88] Ando A, Yang A. Blockade of nitric-oxide synthase reduces choroidal neovascularization. Molecular Pharmacology. 2002;**62**:539

[89] Pechánová O, Simko F. The role of nitric oxide in the maintenance of vasoactive balance. Physiology Research Academic Science Bohemoslov. 2007;**56**:S7-S16

[90] Monaghan-Benson E, Hartmann J. The role of vascular endothelial growth

factor-induced activation of nadph oxidase in choroidal endothelial cells and choroidal neovascularization. The American Journal of Pathology. 2010;**177**:2091-2102

[91] Ushio-Fukai M. Redox signaling in angiogenesis: Role of NADPH oxidase. Cardiovascular Research. 2006;**71**:226-235

[92] Wang H, Yang Z. Endothelial nadph oxidase 4 mediates vascular endothelial growth factor receptor 2-induced intravitreal neovascularization in a rat model of retinopathy of prematurity. Molecular Vision. 2014;**20**:231-241

[93] Al-Shabrawey M, Rojas M. Role of nadph oxidase in retinal vascular inflammation. Investigative Ophthalmology and Visual Science. 2008;**49**:3239-3244

[94] Meng D, Mei A. NADPH oxidase 4 mediates insulin-stimulated HIF-1 and VEGF expression, and angiogenesis in vitro. PLoS One. 2012;7:e48393

[95] Pouw AE, Greiner MA, Coussa RG, Jiao C, Han IC, Skeie JM, et al. Cell-matrix interactions in the eye: From cornea to choroid. Cell. 2021;**10**(3):687. DOI: 10.3390/cells10030687

[96] Nita M, Strzałka-Mrozik B, Grzybowski A, Mazurek U, Romaniuk W. Age-related macular degeneration and changes in the extracellular matrix. Medical Science Monitor. 2014;**20**:1003-1016. DOI: 10.12659/MSM.889887

[97] Campochiaro PA. Molecular pathogenesis of retinal and choroidal vascular diseases. Progress in Retinal and Eye Research. 2015;**49**:67-81. DOI: 10.1016/j.preteyeres.2015.06.002

[98] Yang M, So KF, Lam WC, Lo ACY. Novel programmed cell death as therapeutic targets in age-related macular degeneration? International Journal of Molecular Sciences. 2020;**21**(19):7279. DOI: 10.3390/ijms21197279

[99] Hanus J, Anderson C. Wang S.RPE necroptosis in response to oxidative stress and in AMD. Ageing Research Reviews. 2015;**24**(Pt B):286-298. DOI: 10.1016/j.arr.2015.09.002

[100] ElShelmani H, Wride MA, Saad T, Rani S, Kelly DJ, Keegan D. The role of deregulated MicroRNAs in age-related macular degeneration pathology. Translational Vision Science & Technology. 2021;**10**(2):12. DOI: 10.1167/tvst.10.2.12

[101] Stockwell BR, Friedmann Angeli JP, Bayir H, Bush AI, Conrad M, Dixon SJ, et al. Ferroptosis: A regulated cell death nexus linking metabolism, redox biology, and disease. Cell. 2017;**171**:273-285

[102] Čolak E, Žorić L, Radosavljević A, Ignjatović S. The Association of Serum Iron-Binding Proteins and the antioxidant parameter levels in age-related macular degeneration. Current Eye Research. 2018;**43**(5):659-665. DOI: 10.1080/02713683.2018.1437452

[103] Yang WS, Stockwell BR. Ferroptosis: Death by lipid peroxidation. Trends Cell Biology. 2016;**26**:165-176

[104] Bayir H, Anthonymuthu TS, Tyurina YY, Patel SJ, Amoscato AA, Lamade AM, et al. Achieving life through death: Redox biology of lipid peroxidation in ferroptosis. Cell Chemical Biology. 2020;**27**:387-408

[105] Green DR. The coming decade of cell death research: Five riddles. Cell. 2019;**177**:1094-1107

[106] Birk M, Baum E, Kouchek Zadeh J, Manicam C, Pfeiffer N, Patzak A, et al.

Angiotensin II induces oxidative stress and endothelial dysfunction in mouse ophthalmic arteries via involvement of AT1 receptors and NOX2. Antioxidants. 2021;**10**:1238

[107] Carbajo-Lozoya J, Lutz S, Feng Y, Kroll J, Hammes HP, Wieland T. Angiotensin II modulates VEGF-driven angiogenesis by opposing effects of type 1 and type 2 receptor stimulation in the microvascular endothelium. Cellular Signalling. 2012;**24**:1261-1269

[108] Kim JH, Yu YS, Cho CS, Kim KW. Blockade of angiotensin II attenuates VEGF-mediated blood—Retinal barrier breakdown in diabetic retinopathy. British Journal of Pharmacology. 2008;**29**:621-628

Chapter 2

Assessment on Oxidative Stress in Animals: From Experimental Models to Animal Production

Gustavo Alberto De La Riva De La Riva,
Luis Adrián Saldaña Trujillo
and Juan Carlos González-Hernández

Abstract

Oxygen is a key element involved in a variety of vital physiological reactions in aerobic organisms, including those produced in the electron transport chain, hydroxylation, and oxygenation. Reactive oxygen species and reactive oxygen nitrogen species (ROS/RONS) are naturally formed as by-products from these previously mentioned processes and reactions involving the O_2 molecules. Under healthy conditions, the harmful effects of ROS/RONS in the organisms are controlled by antioxidants, molecules of enzymatic or non-enzymatic nature, able to prevent, retard, or eliminate oxidative damage. Nevertheless, when ROS/RONS production exceeds the antioxidant capacity of one organism, oxidative stress emerges, leading to the apparition of many diseases, some of which can depict significant losses in the field of animal production. Thereby, looking for increasing animal productivity, procedures to mitigate the effects of oxidative stress on living organisms are tested in laboratory animal models, and the obtained results are used to develop strategies that avoid oxidative stress in farm animals either invertebrates (mollusks and crustacean species) or vertebrates (fish, birds, and mammals). In this chapter, oxidative stress will be addressed from the field of animal health and welfare and its impact on animal production, presenting some strategies, studies conducted, and recent perspectives to mitigate the effects of oxidative stress and improve the productivity indicators in farm animals.

Keywords: oxidative stress, environmental stress, livestock handling, anti-stress procedures, animal production

1. Introduction

Our planet was formed as a condensed mass of cosmic debris about 4.5 billion years ago. It had a primitive ocean, atmosphere, climate, and recurrent cataclysmic events. This hostile and hellish environment eventually became more suitable for the first forms of life, but this was a life without oxygen [1]. In the atmosphere, oxygen appeared primarily as a result of solar radiation disruption of water (H_2O) into hydrogen (H_2) and oxygen (O_2) and, much later, by the photosynthetic process. Consequently, anaerobic life forms have to adapt to using oxygen and/or contend with the potential dangers elicited by oxygen and its metabolic by-products.

Oxygen is a key element involved in a variety of vital physiological reactions in aerobic organisms, including those produced in the electron transport chain, hydroxylation, and oxygenation. Only a very small proportion of oxygen used in cells (about 1%) is transformed into reactive oxygen-nitrogen species (ROS) and reactive oxygen and nitrogen species (RONS), which are primarily a result of essential biochemical pathways associated with aerobic metabolism, which takes place in the mitochondria and peroxisomes [2] and includes not only some radicals such as hydroxyl radical (HO-), peroxyl (ROO-), but also other molecules such as superoxide anions (O_2-) hydrogen peroxide (H_2O_2), reactive aldehydes (ROCH), and nitric oxide (NO). Intending to mitigate the adverse effects caused by ROS/RONS, organisms use antioxidants defense system composed of endogenous molecules and molecules having nutritional nature, able to retard or inhibit the damage produced by free radicals when present in a lower concentration are concerning them [3]. The cellular antioxidant systems composed of several enzymes, including superoxide dismutases (SOD), catalases (CAT), and glutathione peroxidases (GPx), together with other antioxidant molecules such as thioredoxin 2 (TRX2), glutaredoxin 2 (GRX2), cytochrome c oxidase (complex IV), coenzyme Q, ascorbic acid, tocopherol, vitamin E, and carotene, among others [3–5].

Stress is the response reflex reaction exposed by the incapacity of an organism to face its environment. This condition may lead to unfavorable consequences, ranging from discomfort to death: a concise review of the impact of stress on growth, production, reproduction, and disease [6]. When ROS/RONS production exceeds the antioxidant capacity of one organism, oxidative stress emerges. Oxidative stress may be involved in several pathological conditions, including conditions relevant to animal production and the general welfare of individuals [7]. Farm animals usually suffer from the onslaught of oxidative stress as a result of the unsuitable enclosure and overpopulation of stables, nutritional and health conditions, and their handling according to the production cycle established for each species [8–10].

The aim of this chapter is to present some studies on oxidative stress carried out in experimental living models, making emphasis on how experimental data gained in these research studies can be applied to farm animals through some ways to measure and nutritional strategies to reduce the adverse effects that oxidative stress produces on the health and physiological development of wild and farm animals.

2. Generation of oxidative stress

We can group the stressors in three categories: endogenous and pathogenic illness and external. ROS/RONS production is a key mechanism involved in the damage caused by different diseases and infections provoked by both pathogenic and no-pathogenic organisms.

2.1 Endogenous sources

The ROS/RONS has both enzymatic and non-enzymatic nature. Numerous biochemical processes also generate those types of reactive species including such acting as intercellular signals with remarkable incidence in specific biological processes and functions [11, 12]. In the cell, mitochondria have the ability to produce ATP through respiration and mitochondrial oxidative phosphorylation, the most important source of ATP [13]. The mammalian cytochrome P450 CYP-dependent ROS/RONS electron transport system and the mitochondrial electron transport chain are the most important source of ROS/RONS. In mitochondria, the formation of such reactive substances by the activity of NADPH oxidase (NOX) enzymes can also be possible, generating superoxide anion (O_2^-) from cytoplasmic NADPH [14]. In endoplasmic reticulum, cytochrome P450 (CYP) enzymes play an important role in the activation of oxygen, generating $O_2{\bullet}^-$ while catalyzing NADPH-dependent reactions in electronic transport chain. Cytochrome P450 is the main way of xenobiotic detoxification, including drugs, medications, chemical products, alcohol, aromatic substances, pesticides, and other industrial byproducts. Cytochrome P450 houses enzymes catalyzing the oxygenation of organic substrates and the simultaneous reduction of molecular oxygen, and it plays a key role in the reduction of O_2, hydroperoxides, arene, N-oxides, azido and azodic compounds, halogen, nitrogenic compounds, hydroxylamines, and other xenobiotics. Some intermediaries of the processes mentioned above ROS/RONS initiate lipid peroxidation and directly damage the DNA and the cellular membranes [15]. Peroxisomes are the important generators of ROS/RONS. The main function of peroxisomes is to break down long fatty acid chains through beta-oxidation and synthesize necessary phospholipids, such as plasminogen, which in vertebrates are critical for correct function of the brain, lung, and heart. Furthermore, this subcellular structure aids certain enzymes with energy metabolism in many eukaryotic cells as well with cholesterol synthesis in animals [16]. The major sources of ROS/RONS in central nervous system are the microglia cells, which support neuronal survival through the secretion of growth factors and anti-inflammatory cytokines. Activated microglia cells produce various pro-inflammatory mediators, nitric oxide, and ROS/RONS; however, their prolonged activation can cause many neurodegenerative disorders, as it was previously exposed, the phagocytosis is an important ROS/RONS generator, and this process is intrinsically linked to pathogenic infections as a part of immune response [17, 18].

In animals, oxidative balance could be broken by different organic, nutritional, and environmental conditions leading to different pathologies including neurodegenerative, cardiovascular, hepatic, renal, metabolic, and autoimmune diseases, and even behavioral and reproductive disorders [11, 19–21]. Farm animals live usually suffer from the onslaught of oxidative stress due to unsuitable enclosure and overpopulation of stables, nutritional and health condition, and their handling according to the production cycle established for each species.

2.2 Exogenous sources

Exogenous conditions can promote the production of ROS/RONS and induce oxidative stress in the organism. Factors such as illness and pathological process, drugs and medicaments, exposure to radiations, chemical by-products of industry, pollution, and adverse environmental and living conditions induce oxidative stress [22]. ROS/RONS such as hydroxyl radicals and superoxide are critical in the apoptosis of infected cells as it has been demonstrated [23]. Pathogenic bacteria can provoke inflammation and destruction of infected cells and tissues. Experimental data suggest that periodontal disease-induced oxidative stress and inflammation is mediated through purine degradation pathway, a major biochemical source for ROS/RONS production [24]. The connection between bacterial pathogens and unfolded protein response (UPR) evidences that pathogenic bacteria induce oxidative stress in the endoplasmic reticulum. This suggests this is the route to access and gain nutrients from the host, obviating the need to become internalized or inflict irreversible cell damage [25]. Other example is the case with idiopathic pulmonary fibrosis (IPF), a fatal lung disease of unknown origin characterized by chronic and progressive ROS/RONS interstitial pneumonia, which progressively impairs lung function [26]. Oxidative stress induces hypoxic tissue condition and injuries in different cell and tissues [27].

3. Role of ROS/RONS in healthy organisms.

The positive roles of ROS/RONS are widely recognized in signal transduction [28, 29], in the induction of endothelial cell migration and proliferation, apoptosis [30], macromolecule oxidation, structural and functional modulation, regulation of gene expression [31–33], mRNA stability, and methylation-mediated DNA epigenetic factors [33, 34]. Phagocytosis and other important aspects of immune defense generate ROS/RONS [35, 36].

Determination of the ROS/RONS sources and the etiology of oxidative stress are important for their implication in human health. The application of specific therapies and clinical treatments to reestablish oxidant/antioxidant balance in the organism is an important part of recovery of health from a convalescent diseased condition. The studies in different animal models definitively contribute to this goal, but the results can be also suitable to extend and apply to other animal species including wild, pets, and farm animal to mitigate the effects of oxidative stress. For farm animals, live in stable conditions and cycles of handles led to the periodical changes in their habitat conditions causing oxidative stress with adverse consequences in the productive and economic indicators of livestock activity [9, 37–39]. The results obtained and the perspectives are encouraging with the consequent improvement of their health, optimal development, and quality of life.

3.1 Cellular models

Escherichia coli has largely been used as model microorganism in biological research studies including the generation of ROS/RONS and oxidative stress [40]. Specific protein groups were identified in *E. coli* due their antioxidative properties including molybdenum-binding enzymes and fimbriae assembly proteins for understanding important features of the structural proteome to enable modeling of different stress responses [41]. To mitigate the damage of oxidative stress, different stress response regulons are activated in bacteria, depending on the type of stressor [42].

Yeasts such as *Saccharomyces cerevisiae* and *Schizosaccharomyces pombe* are model species for basic studies in cell biology, based on powerful genetic tools that allow gene disruption and phenotypic analyses together with more sophisticated functional screens [43, 44]. Yeast remains single a cell but eukaryotic organism, ant it is evolutionarily more advanced than *E. coli* harboring more complex cellular structures and biological processes. One of the themes in which yeast studies have provided considerable information is the cellular response against oxidative stress and the defense functions involved in such response [45, 46]. Yeast allows study of oxidative stress and effects in different cell macromolecules, which has been related to a number of human diseases. Although the studies in *E. coli* offer a general spectrum of oxidative stress and how it impacts in cell physiology, the studies in yeast bring the opportunity to see the oxidative stress in eukaryotic cells, where the physiological and molecular processes are more complex.

To elucidate the individual effects of numerous oxidative stress-inducing agents that surround the all living beings and test different strategies to manage the adverse consequence of oxidative stress and restore the oxidant/antioxidant balance, frequently it is necessary to test and elucidate the influence and role of oxidative stress by external agents on general cellular processes. Different unicellular lines can be used also, but when it is necessary to study these effects in specific differentiated cells, these can be taken from animal tissues and organs to carry out transient tests with them [47–50]. Primary cell culture consists of a temporary or transient culture of differentiated cells removed from tissues and plated in appropriate media and provides an opportunity to assess such effects, as well as testing some methods to assess oxidative stress conditions. Some of the most used animal and human culture cells are blood erythrocytes, leucocytes, leukemia cells, hepatocytes, human hepatoma, pheochromocytoma cells, colo-adeno carcinoma cells, among others [51–53]. The tests in primary cell cultures are very convenient because the methodology is simple and well established, and also, the tests are usually fast and low cost. But primary cell cultures are different from those isolated directly from tissue, and originally these cells were parts of organs and tissue in living organisms, in contact with other differentiated cell and constant and active interaction with them. Without such interaction, the isolated cells are very vulnerable to many agents, including those promoting oxidative stress. The mentioned facts imply that often the results obtained in culture cell could differ with tests performed in living animal models and encouraging results in cellular lines became disappointing in when tried *in vivo* [51, 52]. Cellular primary culture and cellular lines are exposed to high oxygen concentrations so the balance between oxidative processes generating RONS and antioxidant cellular defenses is altered in favor of oxidative stress. In addition, the fragile stability of cultured cells is also present in the conservation of original characteristic (genotype and phenotype) due the high mutation rate in cellular lines [51, 54, 55].

4. Experimental model and farm animal

4.1 Invertebrates

Invertebrates have been valuable research models in the discovery of many biological principles owing to the numerous advantages they provide. During the life cycle, many in pathogen-rich environments manage harsh weathers, exposed to a number of chemical compounds, and they are well adapted to both terrestrial and marine ecosystems. Their remarkable ability to successfully face enormous oxidative stress generated in all these circumstances makes them attractive models for research. In addition to mollusks, one of the most recurring invertebrate animal models is *Caenorhabditis elegans*, a free-living nematode that lives in temperate soil environments, and *Drosophila melanogaster*, a widely used model organism, has provided insight into eukaryotic genetics and human disease.

4.1.1 Mollusks

Mollusks (*Mollusca*) are the second largest phylum of invertebrate animals by the total number of species. The survival of aquatic mollusks depends on their ability to sense and respond appropriately to biotic and abiotic changes in their habitat. The studies and changes in their immune responses can also serve as indicators of changes in ocean environments. Therefore, studies into understanding new factors in their immune systems may aid new biomarker discovery and are of considerable value. In general, these studies were focused on ecotoxicological aspects in mussels such as Mediterranean mussel (*Mytilus galloprovincialis*), Atlantic ribbed Mussel (*Geukensia demissa*), Rooved carpet shell *Ruditapes decussatus*, Blue mussel (*Mytilus edulis*), Soft Shell Clam (*Mya arenaria*), Eastern Oyster (*Crassostrea virginica*), and Atlantic Jack-knife Clam (*Ensis leei*) among others [56–58]. Aquatic and terrestrial mollusks are exposed to metallic polluted water and substrates with contaminants show a reduction of lower rates of reduced glutathione (GSH) when lived in a water or substrate deliberately contaminated with copper and zinc [59]. But at slightly lower levels, effect has been reported in terrestrial mollusks due to the fact that in terrestrial mollusks the rate of ingestion is slower. This behavior induces lipid peroxidation that is well correlated with the reduction in the whole antioxidative enzymes [60]. Not only water and soil pollution induces oxidative stress affecting oxidant/antioxidant balance, abiotic environmental conditions due to the seasonal variation in water temperature, salinity, the rate of dissolved oxygen, the nutritional value and quantity of food available [56, 60], UV radiation, pH, and dissolved O_2, [61]. Bivalve mollusks from freshwater to marine ecosystems are suspension feeders and have been widely studied in ecotoxicology as indicator species for monitoring of environmental perturbations. They have highly efficient cellular internalization system for external compound and micro- and nanoparticles able to modify and disturb vital physiological functions including intracellular digestion and cellular immunity [56, 62].

Mollusk communities are getting endangered by oxidative stress and different mollusks have been used as biological models to study the effect of oxidative stress and the antioxidant mechanisms to alleviate the relating damage. The most important studies on OS were focused on the impact of pollution-induced OS generating physiological and morphological modifications due to carcinogenic and toxic nature of many chemical compounds. The most frequent consequences are neoplasies as blood neoplasm, hematopoietic neoplasm, disseminated neoplasies, hemic neoplasies, leukemia, and proliferate cellular disorder [63]. Many compounds such as polyaromatic hydrocarbons (PAH), polychlorinated biphenyls (PCB), and heavy metals (Pb, V, Cr, Ni, Cu, Mn, Zn, Cd) are implicated in the induction of the production of RONS although many specific aspects of the mechanism remain unknown [64]. The alteration in catalases, superoxide dismutase, glutathione peroxidases, glutathione S-transferase, and other enzymes was observed as well as higher rates of reduced glutathione. These alterations of OS indicators were found mainly in gills and in the digestive gland, the most exposed organs, indicating the activation of antioxidant defenses [56].

4.1.2 Shrimps

In the last two decades, there has been an increase in crustacean aquaculture production, reaching approximately 9.3 million tons in 2018. The increase in production output is primarily due to aquaculture intensification, such as improved feed efficiency, feed formula, and feeding management techniques and technological improvements, with considerable impact on protein efficiency and nitrogen retention. Intensive aquaculture is associated with several stressing factors, including increased stress resulting from high stocking density, which can cause high mortality. To face these stress conditions, white shrimp (*Litopenaeu vannamei*) and shrimp in general display an innate immune system that is characterized by defense mechanisms based on cellular and humoral reactions that are mainly related to the hemolymph. Hemocytes have been identified as immune defense components that exhibit high phagocytic activity [65]. The modulation of immunological parameters in response to adverse conditions is an important indicator of shrimp health status. High stocking density conditions require the adaptation of shrimp to lower rate of dissolved oxygen levels, reduced space, and increased interactions with other organisms, which lead to increased energy consumption for the maintenance of physiological conditions These facts have direct effect on cellular hemocyte composition [66, 67].

Fatty acids are essential for the constitution of the body and maintenance of vital metabolic activities. The n-3 highly unsaturated fatty acids (n3-HUFAs), such as EPA and DHA, are important components of the phospholipids of the cell membrane, determining their fluidity, and they are involved in other functions, such as the reproductive and immune functions of aquatic organisms [68]. Crustaceans have difficulty in converting polyunsaturated fatty acids into highly unsaturated fatty acids (HUFAs) through the elongation of carbon chains and the desaturation of fatty acids [69]. Moreover, it has been determined that high levels of n3-HUFAs can be easily oxidized, leading to oxidative stress [70, 71]. Determination of the optimum amount of fatty acids required in shrimp diet is very important. A decrease in growth and an increase in mortality at dietary lipid levels greater than 10% have been reported previously [72]. Additionally, it has been observed that high lipid levels in feed can affect the lipid composition of shrimp tissues, resulting in high concentrations of lipids in the digestive gland and muscles [73]. Increases in respiratory burst capacity, total hemocyte count (THC), and CAT and GPx activities are well correlated with increasing dietary lipid levels, suggesting that the optimum lipid supplementation level is between 10% and 12% [74].

The best performance with respect to zootechnical and immunological parameters was observed in shrimp when they were fed with high-carbohydrate diets, and the highest growth rate and hemocyte number obtained in shrimp fed diets with approximately 40% carbohydrate under different stress conditions [75]. Providing shrimps with diets rich in nutrients (e.g., proteins, lipids, carbohydrates, vitamins, minerals, and additives) is a suitable approach for improving performance of animals, faster development, and increases in productivity as well as more resistance against oxidative stress [76]. Due its economic importance, the different dietary carbohydrate/protein and lipid/protein ratios on the growth performance, protein efficiency, nitrogen retention, and immunological and oxidative stress status of white shrimp are cultured under an intensive system [77].

4.2 Vertebrates

4.2.1 Fishes

Fish, like all living organisms, suffer from the consequences of oxidative stress caused by the generation of reactive ROS/RONS species induced by different factors such as diseases and the processes associated with them, capture, transport, and handling of fish, hypoxia, environmental temperature, inadequate salinity of the waters, both by chemical nature of dissolved salts and per their concentration, malnutrition, pollution by different types of contaminants. The sea and freshwaters are prone to contamination by flowing into both surface and underground water reservoirs, through rain runoff and natural or artificial drainage lines to rivers, lakes, and seas. Among these contaminants, we can highlight those that induce the generation of ROS/RONS at levels that overload the antioxidant defense of the organism altering the normal morphological and physiological functioning of the organs and tissues [78]. The first important indicators to test pollution and environmental affectations in water-inducing oxidative stress are the deterioration of DO level (dissolved oxygen in water), pH, temperature, and other basic parameters of aquatic environment [79]. Described alterations led significant affectations on fish morphology, growth rates, and reproductive functions with adverse combinational and cumulative affectation in quantity and quality of fish production.

Zebrafish is a valuable and one of most versatile experimental animal models particularly in Ref. to fish biological and genetic studies due its small size, rapid development, and genetic and biochemical homology with higher vertebrates. It has been used for drug screening, determination of bioactivity, toxicity, and collateral side effects of novel drug candidates, but it is also a useful experimental model to study oxidative stress-linked disorders [80]. Zebrafish larvae were used to evaluate the antioxidant activity of 15 commercially available flavonoids against UV-induced phototoxicity, along with the computational quantitative structure–activity relationships (QSAR) method to investigate the correlations between the observed biological activities and the physicochemical properties of the different compounds. Among these compounds, chrysin and morin showed higher ROS-scavenging rates (99 and 101%, respectively) and lower toxicity (LD50 > 100 ppm) [81]. Flavonoids are the most abundant polyphenols in our diet and the most studied ones. They are characterized by a common benzo-γ-pyrone moiety, variously substituted with hydroxyl and methoxyl groups. Based on their chemical structures, flavonoids may be divided into six major subclasses: flavonols, flavones, flavanones, flavanols, anthocyanins, and isoflavones [82–84].

The studies on oxidative stress carried on this model help to understand the effect of oxidative stress in ecological and economic fields applied to both wild and aquaculture species. About 60 million people around the world are involved in the primary sector of capture fisheries and aquaculture, and this economic field involved in capture, fish handling, processing, and selling, making this activity an important source of human food and animal feed [85, 86]. Industrial fishing is highly developed sector but capture fishing pressure faces growing environmental problems and other operative adverse factors, as overexploitation of marine and freshwaters species, climate change, and toxic pollution. Recent data from the Food and Agriculture Organization of the United Nations (FAO) indicate that the share of marine fish stocks within biologically sustainable levels is declining [87]. Worldwide, the most important fish species used in fish aquaculture are carp (*Cyprinus carpio* and *Cyprinus* ssp), salmon (*Salmo* ssp), tilapia (*Oreochromis* spp. y *Tilapi*a sp), catfish (*Silurus* ssp), and trout (*Oncorhynchus*) [87].

Different taxa and different species of the same taxa of fishes exhibit differential tolerance to oxidative stress [88]. The oxidative stress can affect reproduction, fecundity, and genetic degeneration in progeny, leading to the extinction of some of the fish species [89]. Some chemicals such as duroquinone-inducing oxidative stress affect the spermatozoa of common carp (*C. carpio*) and reproductive damages were also reported in fathead minnow (*Pimephales promela*). Inadequate aquatic environment greatly affects the growth and development of fish affecting food chain and entire trophic communities, causing significant losses to traditional fishing production and that is carried out in a semi-intensive and intensive way in aquaculture farms [89].

4.2.2 Poultry

Poultry in commercial settings is exposed to a range of stressors. A growing body of information clearly indicates that excess ROS/RONS production is the major detrimental consequence stressor inducing oxidative stress in poultry production [90]. In addition, other environmental stressors such as NH_3 pathogenic infections are considered also as risk factors for oxidative stress in all animals that are raised under the intensive mode of livestock production [91]. Farm chickens live in a pro-oxidative environment, due to the high population densities, fluctuations in temperature, and high levels of ammonia due to bird droppings and handling of animals during the entire productive cycle, affecting poultry health and production parameters. The genetic selection on fast growth rates and complexion, selected under commercial criteria with lean and large breast, makes farm chickens and other farm birds more susceptible to oxidative stress [92]. The economic impact of oxidative stress farm bird industry is enormous. Poultry is one of the most and fastest-growing animal production sectors and has a substantial contribution to food security due to the accessibility and nutritional value of its products, particularly meat and eggs, for the majority of the world's population. This livestock sector exhibits one of highest rate of growth and technological innovation and improvement and is listed among the largest world agricultural commodities. The poultry sector, which has a market value of $ 310.7 billion in 2020, is expected to grow to $ 322.55 billion in 2021 and record at compound annual growth rate (CAGR) of 3.8%. The market is expected to hit $ 422.97 billion in 2025 [93].

The main source of RONS in chicken muscles is the leakage of electrons from the respiratory chain in mitochondria during the reduction of molecular oxygen to water process observed in all aerobic organisms, as in other studied animal models [14]. Among animal source foods, poultry meat has also been recognized as a material highly sensitive to oxidative processes owing to the high unsaturation degree of the muscle lipids and the susceptibility of meat to undergo oxidative reactions involves many other endogenous (i.e., antioxidant enzymes) and external factors including environmental conditions during the period of growth and fattening, for animals destined for meat; and of growth and productive period of eggs. According to the literature reports, heat and diet are the most remarkable means of oxidative stress in domestic birds. These adverse factors may lead to biological damage, serious health disorders, lower growth rates, and, hence, economic losses [92]. Supplementation of feed using antioxidant dietary has been proposed to alleviate the oxidative stress of poultry produced in hot climates although low temperature also decreases the level of antioxidant activity lowering the antioxidant defenses in the animal in different tissues and organs [94, 95].

Oxidative stress can affect important organs and tissues, including liver, kidney, reproductive organs, and immune function. The liver plays a crucial in avian species, compared with the majority of most mammalians, including rodents, ruminants, swines, equines. Liver is the most important organ for "*de novo*" lipogenesis in birds and about 90% of fatty acids that are synthesized in liver, where the rest, the minor proportion, is contributed from adipose tissue. This fact is even more significant in commercial poultry due to the fact that dietary fat content is relatively low [96]. Fatty acids synthesized by the liver are transformed, by the biochemical route of low-density lipoproteins (VLDL), which act as an energy source to other tissues for immediate use or storage. The critical role of avian liver in lipid metabolism is also highlighted during egg production, which demands a shift of hepatic lipids to the yolk to nourish the embryo [96, 97]. RONS can oxidize and damage cellular proteins and lipids, but are normally kept in check by the liver's robust system of antioxidants and antioxidative enzymes that quickly neutralize excess RONS to maintain the redox balance [90, 98]. It has been observed that in the liver cells the mitochondrial function and other complex activity in low-feed efficient broilers. This fact is associated with higher oxidative stress and differential protein expression [99].

The levels of essential microelements in poultry feed must be high enough to satisfy the birds' requirements and, at the same time, low enough to ensure the safety of both animal feed and meat and eggs for human nutrition. The essential role of vanadium (V), chromium (Cr), and nickel (Ni) in poultry nutrition is still under investigation, while their toxicity was well established a long time ago. Vanadium plays a role in lipid metabolism and its deficiency in feed can be associated with decreased levels of blood and bone iron, which can result in abnormal bone development. A range of experiments indicated the crucial role of trivalent Cr (Cr^3+) in the maintenance and regulation of blood glucose levels (*via* the glucose tolerance factor – GTF). Dietary Cr^3+ improves insulin effectiveness by enhancing its binding to receptors and the sensitivity of the target cells [100]. In chickens, insulin promotes high rates of protein synthesis and the better amino acid transport and lower rates of protein degradation. This fact increases the weight of pectoral muscles and the meat of these broilers contained less fat and cholesterol causes disorders of carbohydrate and protein metabolism, reductions in insulin sensitivity in the peripheral tissues, and decreases in growth rate [100]. The dietary addition of Cr propionate improved feed efficiency and decreased mortality [101].

The essential function of Ni has been established in rats, poultry, swine, goats, and sheep. Ni plays a role either as a structural component in specific metalloenzymes (urease, hydrogenase). This element acts also as a cofactor in facilitating intestinal absorption of ferric ions. Experiments have suggested its importance in cardiovascular pathology; that is, elevated serum Ni concentrations are observed in patients with acute myocardial infarction and stroke [102]. Ni ions have a high affinity for proteins and amino acids and cause protein oxidation in cells. Its binding to chromatin in somatic cells can result in oxidative and structural damage to cell proteins. Currently, eight enzymes containing Ni have been identified [103].

Selenium is an essential micronutrient and well antioxidant naturally found in soil, water, and some foods. Selenium compounds in trace quantities are indispensable for proper physiological functioning of vertebrate organisms. The beneficial effects of selenium are related to the selenoproteins, playing relevant role in the many physiological functions, including endocrine, muscular, cardiovascular, nervous, reproductive, antioxidant, and immune functions [10, 104, 105]. Selenium compounds improve immune responses modulating the production of certain cytokines secreted by cells of the immune system and enhancing the resistance of the immune cells to the oxidative stress. Selenium supplementation had inhibitory effects on tumor necrosis factor alpha (αTNF) levels in heat-stressed broiler chicks, but the details are not completely elucidated and are in continuous investigation [10, 106]. These mentioned microelements in inorganic formulations are suitable for consumption in all animals as dietary supplement, but in trace well-established amounts, above which they are harmful to the health causing severe intoxication and inducing oxidative stress. The damage by these stressors has been observed, with particularly severity, in liver and kidneys and gonads, suggesting that these organs are greatly vulnerable to metal intoxication with the logical affectation of hepatic, renal, and reproductive functions [55, 107]. Environmental problems also affect the farm animals in intensive husbandry, and ammonia (NH_3), a severe air pollutant, is an important factor for the formation of secondary particles in the heavy haze pollution [108, 109]. It is known that high concentrations of atmospheric ammonia induce alterations in the hepatic proteome of broilers [91, 110].

4.2.3 Rodents

Rodents are mammals of the order *Rodentia*, which are characterized by a single pair of continuously growing incisors in each of the upper and lower jaws. About 40% of all mammal species are rodents. Difference has been extensively served as experimental models: guinea pigs (*Cavia porcellus*), mice (*Mus musculus*), rats (*Rattus norvegicus domestica*) syrian hamster (*Mesocricetus auratus*), or rabbits (*Oryctolagus cuniculus*). It has been firmly established that rats, mice, and humans each have approximately 30,000 genes of which approximately 95% are shared by all three species [111]. Other advantages for using rodents as animal model for scientific research are relatively small and require little space or resources to maintain, have short gestation times and large numbers of offspring, and have quite rapid development to adulthood and relatively short life cycle spans. Mice have a gestation period of about 3 weeks, a fourth week period of weaned, and reach sexual maturity by 5–6 weeks of age. This short life cycle allows large numbers of mice to be generated for studies fairly quickly. However, in many areas of research rats are preferred, including cardiovascular research, behavioral studies, and toxicology.

The rabbit (*O. cuniculus*) is phylogenetically closer to primates than other rodents [112] and is large enough to permit non-lethal monitoring of physiological changes. The rabbit is also standard laboratory animal in biomedical research, and transgenic rabbits are used as animal models for a variety of human genetic and infection diseases. It is routine, the use of rabbits includes antibody production, development of novel surgical techniques, physiological clinical methods, and toxicological studies for the testing of new drugs and development of new medical treatments. There are a great number of research studies where mice and rabbits, including genetically transformed animals, were used in many important topics, including lipoprotein, atherosclerosis, cardiovascular research, and hypertrophic cardiomyopathy. Some of these mutants haven conceived to constitutively develop oxidative stress, which in some cases lead to the status for the spontaneous development of tumors. One of the most remarkable cases is the mutants where gene encoding for well-known tumor suppressor p53 is knocked out. In humans around 50% of cancer studied cases have shown this suppressor gene mutated [113]. The p53 gene is induced and expressed in response to different type of stresses, including oxidative stress. This is an example when genomic stability and integrity are endangered *via* DNA damage under oxidative stress. Many genes, acting in basic functions as a transcription factor p53 gene, are targets of oxidative stress. There many other genes encoding for regulatory proteins or functional RNA, regulated either positively or negatively, are targets for oxidative stress attack. These target genes are usually involved in various cellular processes converging into genome stability maintenance such as cell cycle arrest, senescence, apoptosis, and DNA repair, suggesting a possible reason to explain the tendency in those mutants to develop cancerous tumors [114].

Mouse is the most widely used model in biochemical research, and mice provide ideal animal models for biomedical research and comparative medicine studies because they have many similarities to humans in terms of anatomy and physiology. Oxidative stress mutant mouse models were either specifically designed to assess the oxidative role of candidate molecule or shown to be oxidative models "by chance." The models genetically modified mutant mice are transgenic animals where a deregulation of redox status has been modified by genetic transformation [115, 116]. One of these mutant mouse models lacks the transcription factor Nrf2, which is very important in the regulation of expression of phase II-antioxidant enzymes genes. Each mutant has a knock-out of one of the components of the cellular antioxidative defense disabling an adequate full response to oxidative stress. Some mutants also are susceptible to neoplastic tumors induced by increased ROS/RONS level. This applies to superoxide dismutase 1 (SOD1) and 2 (SOD2), and such animals develop liver cancer and are conducive to the development of cancerous tumors. The case for peroxiredoxin 1 (PRDX1)-deficient animals that die from malignant cancers after surviving hemolytic anemia, and for OGG1-deficient mice in which a defect in the elimination of oxidative-damaged DNA facilitates lung tumor development [117]. Studies in mice and other rodents can be preliminary steps not only to applied successfully assayed prophylactics, drugs, and treatments to human health but also to wild and farmer mammalian animals. Some rodents are not only experimental models, but at the same time they are animals of economic interest and the basis of an important food industry, that is, the case of rabbits, for example. This is an important fact that allows extending obtained results on the causes and effects of oxidative stress not only under animal husbandry conditions but also under production conditions, providing a robust basis for applying appropriate strategies in other farm animals such as pigs and cattle.

4.2.4 Swines

Swines (*Sus domesticus*) are an important species in livestock production. Total global pork exports for 2021 are expected to reach 11.8 million tons, mainly with China as a lider, followed by the EU and United States, and the global pork meat market size is constant growth and the projection by 2027 is expected to reach US $257,874.5 million [118]. Many breeds of pig exist, with different colors, shapes, and sizes. According to The Livestock Conservancy, as of 2016, three breeds of pig are critically rare (having a global population of fewer than 2000 [119].

As other farm animal, pigs are exposed to various types of stressors during their life cycle: dietary, social, environmental stress, but also metabolic stress through high performance in intensive livestock farm. Some stages are especially critical and can induce ROS/RONS and induce oxidative stress. In the weaning phase, piglets often show growth depression and are vulnerable more susceptible to diseases, a phenomenon known as post-weaning stress syndrome [120]. Pigs suffer oxidative stress during late pregnancy, high lactation, and weaning. Weaning is one of the most stressful stages in piglets that results in intestinal, immunological, and behavioral changes and animals became more vulnerable to different disease. During this period, pigs are subjected to a number of stressors, including abrupt separation from the sow, transportation and handling stress, a different food source, social hierarchy stress, co-mingling with pigs from other litters, a different physical environment (room, building, farm, water supply, etc.), increased exposure to pathogens, and dietary or environmental antigens. The piglet must adapt to all of these stressors rapidly to be productive and efficient. If the incidence of the different stressors in this stage is too great for the pig, it can lead to poor performance and increased mortality due the oxidative stress. When the piglet is weaned, their young organisms must be abruptly adapted from highly digestible and palatable liquid milk from its mother that is equally spaced throughout the day to a solid dry diet that is less digestible and palatable. As a consequence, feed intake is usually reduced initially after weaning and the piglet becomes malnourished with reduced transient growth rate [121]. In addition, we must point out the trauma that the definitive separation from their mothers causes to young pigs, which also stresses them. Weaning induces both transient and long-lasting modifications of absorptive, secretory, and barrier properties of piglet intestine a well as modification in absorptive, secretory, and barrier functions of the intestine. In the moment piglet stops feeding on breast milk, begins a period of adaptation to many changes stressing their routine normal behavior: the separation of mother and the change in their routine feed. This is a period of variable duration, a period of reduced feed intake. It is considered that by the end of the first post-weaning week, metabolizable energy intake is reduced to about 60–70% of pre-weaning milk intake, and it takes about 2 weeks to complete recovery to the pre-weaning energy intake values [121]. Low-feed intake predisposes the pig to intestinal barrier dysfunction, which is often accompanied by intestinal inflammation and negatively affects villus height and crypt depth.

Environmental problems also affect the farm animals in intensive husbandry, and ammonia (NH_3), a severe air pollutant, is an important factor for the formation of secondary particles in the heavy haze pollution [108, 109]. However, there were a few studies on the effects of NH_3 on pigs but for sure the high-density feeding in livestock houses caused air pollution, and the main by gas pollutant is NH_3 resulting from microbial decomposition of nitrogen-containing organic matter. Some studies have indicated that oxidative stress is one of the toxicity mechanisms of NH_3 [122]. When pigs were exposed to 100 ppm, NH_3 during 6 days the pigs reduced food intake and lost weight and increases up to 100 and 150 ppm provoke acute inflammatory reaction. The effect of ammonia is observed in tracheal epithelium, and this organ is an important barrier between internal organ and tissues and the environment. Ammonia can cause dramatic changes when inhaled, including smooth muscle hyperplasia, pulmonary fibrosis, basal layer thickening, cell composition changes, and inflammatory cell infiltration, loss of cilia or the production of more mucus covered on the basal layer of tracheal cilia. The increased mucus secretion can lead to the development of chronic respiratory diseases including chronic obstructive pulmonary disease (COPD) and asthma [123]. NH_3 exposure leads to loss of cilia or the production of more mucus covered on the basal layer of tracheal tissue as revealed by microscopy observation [122]. The epithelial tissue of trachea forms a cellular barrier limiting the internal organism with the environment. The attack to this barrier by external irritant and harmful substances, particularly gases, combining with other stressor and pathological conditions, alters the function of respiratory structure inducing dramatic chance and muscle hyperplasia, pulmonary fibrosis, damages to the respiratory tissues, inflammation, increase of secretion (mucus), and obstructive respiratory diseases and asthma. All these pathological conditions are accompanied by oxidative stress. Oxidative stress indexes such as superoxide dismutase (SOD), reduced glutathione (GSH), reduced glutathione peroxidase (GSH-Px), and malondialdehyde (MDA) in the tracheal tissue allow the inhalation of ammonia with the damage-producing oxidative stress. The activities of SOD and GSH-Px and the level of GSH in the trachea tissue were significantly decreased compared with the control group, when animals were exposed. On the contrary, MDA content in the trachea tissues of pigs exposed to NH_3 was significantly higher than that in the control group [122]. Transportation of animals is important economic activity in many of the livestock production but is an important stressor affecting many of the farm animals but the transport of animal is also an important stressor that can induce the generation of ROS/RONS and drive the affecting organisms to oxidative stress provoked morbidity, mortality, carcass trim loss, and undesirable meat characteristics [124]. Due to the relevance of this topic within the cattle sector, we will deal with it in the section corresponding to cows.

4.2.5 Cows

Cattle (*Bos taurus*) are large domesticated bovines and the most widespread species of the genus *Bos* [125]. In dairy cows, oxidative stress has a negative impact on immune and reproductive functions: increased mastitis frequency and higher somatic cell counts in milk, decreased fertility, increased embryo mortality, post-partum retained placenta, and early calving. Animal metabolic and physiological traits respond quite differently to different oxidative stress factors and management strategies, but the specific cellular mechanisms of action are not often described, although our observation points out that in general, the mechanism is the same as already was described but particularities derived from intrinsic physiology of each animal species and the characteristics of handling with them during the entire productive cycle. Some oxidative stress factors considered and described are linked with the microbial pathogenic cells inducing mastitis, metritis, and degenerative diseases. In dairy cows, feeding management is based on the use of different concentrates and forage ratios that certainly have effect on some expressed genes associated with oxidative stress pathways. In cows, microbial infections are one of the causes of oxidative stress. Microbial activities cause driver of degenerating oxidative stress diseases. Mastitis and metritis are some of the diseases that have been correlated with other pathogens as: *Streptococcus dysgalactiae, Streptococcus uberis, E. coli,* and *Klebsiella, non-aureus Staphylococci* (NAS), *Staphylococcus aureus,* and *Enterococcus spp.* have been correlated with mastitis occurrence and with oxidative Stress [126]. Mastitis is an intra-mammary infection driven by host-pathogen interactions that cause severe economic losses, including decrease in milk production and quality, premature culling, lower conception rates, and treatment costs in dairy cattle. Therefore, host-microbial interactions have been studied, aiming to optimize the lactating performance of cows [127] and depending on the pathogenic microbes' primary reservoir and mode of transmission. Mastitis has been categorized into contagious and environmental forms. Some detrimental microbes such as *S. dysgalactiae*, *S. uberis*, *E. coli*, and *Klebsiella, non-aureus Staphylococci* (NAS), *S. aureus*, and *Enterococcus* spp. have been well correlated with mastitis occurrence, but unfortunately, the mechanisms underpinning host-microbial interactions inducing mastitis are unclear [128], but probably it is related to a drop in the immune defenses of the animal and what favors the opportunistic infection of these pathogens. The pathogenic invasion mainly occurs from environmental microbes through the mammary teat canal opening in the direction of the tissues, and then to the epithelium cells of the duct, causing inflammation and development of granulation tissues that finally appear as polypoid swelling. Pathogens in the epithelium that remain undestroyed by neutrophils will cause edema, leading to vacuolization and desquamated condition of epithelial acini. The pathogenic bacteria undergo a rapid multiplication, leading to reductions in healthy milk secretory tissue, scanty milk with blood traces as well as gangrene and thrombosis damaging the udder tissue, which can potentially lead to toxemia and death in acute cases. Mastitis milk is source of zoonosis-like tuberculosis, brucellosis, and gastroenteritis. A number of previous studies analyzed the microbiome of mastitis cows compared to healthy ones, indicating a lower microbial diversity in mastitis bovine milk, quantified by a lower Shannon Diversity Index, also called the Shannon-Wiener Index, which is a method to evaluate the diversity of species in a community. The correlation analysis between milk metabolites of the intra-mammary infected cows and their microbial populations has been described [127]. In the case of Metritis, a uterine wall inflammation also is associated by an increasing in n *Fusobacterium, Bacteroides*, and *Porphyromonas* has been associated with metritis [128, 129], while a reduction in the same bacteria has been associated with the antibiotic treatment of this disease. From published data, we can say that both diseases and associated microflora induce oxidative stress.

The increases in ROS/RONS species can provoke lipid peroxidation and other cell disorders including the activation of apoptosis so it is necessary to maintain a robust antioxidant defense to maintain the functional status of the organisms. Dairy cows need an adequate intake of coarse fiber to ensure proper rumination, saliva processing, and rumen buffering; it is recommended that at least 40% of the feed particles containing dietary ingredients in total mixed rations (TMR) for dairy cows be larger than 8 mm [130]. However, commercial cow lactation diets are typically high in concentrate to maximize nutrient intake and milk production efficiency, resulting in diets with low or moderate physically effective neutral detergent fiber (PeNDF) [131]. However, high-concentrate diets are likely to result in metabolic and systemic dysfunction, leading to a rise in the concentration of ruminal volatile fatty acids and a corresponding decrease in ruminal pH [132]. Low pH causes a lysis of rumen microbes and release of endotoxins lipopolysaccharide (LPS) from Gram-negative bacteria, and increases the permeability of the rumen barrier in cows but also in other ruminants. LPS in the rumen fluid is absorbed mostly by the rumen wall, then travels through the ruminal veins to the liver via the portal vein [133], and subsequently reaches the mammary gland, triggering inflammatory responses that result in decreased production. In some tested dairy cattle fed with high concentrate, the oxidant and antioxidant biomarkers such as LPS concentration in the rumen fluid, hepatic vein plasma, portal vein plasma, and jugular vein plasma were higher than in cattle fed with low concentrate [126].

Malondialdehyde (MDA) is a well-accepted and used biomarker to determinate lipid peroxidation and SOD activities. MDA is increased in high concentrate feed formulations for cow, and it is accompanied with a reduction of total antioxidant capacity (T-AOC), GPx, and CAT activity. Regarding the composition and structure of milk metabolites, blood metabolites, hormones, and enzymes, a substantial difference was observed in high concentrate compared to low-concentrate cattle. It was reported that albumin and paraoxonase concentrations are inversely related to oxidative stress due to their contribution to the protection of low-density lipoprotein and high-density lipoprotein against lipid peroxidation, along with protein carbonyl and lactoperoxidase. Increased LPS levels in portal and hepatic veins further damage hepatocytes and impair liver function, shown by enhanced TNF receptor-associated factor 6 (TRAF6), p-NF-B, p38 MAPK, IL-1, and serum amyloid A (SAA) levels in the liver [134]. LPS increases the concentration of LPS-binding protein (LBP), serum amyloid A (SAA), and haptoglobin (HP) in peripheral blood during SARA [135]. The high concentrations of LPS can induce the production of ROS/RONS by Kupffer cells and neutrophils. Kupffer cells, also known as stellate macrophages and Kupffer-

Browicz cells, are specialized cells localized in the liver within the lumen of the liver sinusoids and are adhesive to their endothelial cells, which make up the blood vessel walls. Inadequate regulation of ROS accumulation within metabolically active tissues results in oxidative stress. Isoprostanes (IsoP) are also important biomarkers of lipid peroxidation damage because they form when ROS oxidizes arachidonic acid. Indeed, IsoP was found in the blood and milk of dairy cows during periods of high oxidative stress, such as the peripartum and mastitis periods [136].

Handling of cows also derived in oxidative stress increasing the susceptibility to different diseases with increased incidence and severity of disorders during the transition period (around 3-week prepartum late pregnancy through 3-week postpartum). Transition period is characterized by dramatic physiological and immunological changes from stages of prepartum gestation stage to lactation. It is known that during late pregnancy, glucose and amino acid requirements increase in order to support fetal development. Transition cows may display an overt systemic inflammatory response around the time of calving even without signs of microbial infections or other pathologies [137]. Low-grade chronic inflammation in transition cows has adverse effect and can aggravate metabolic stress due the increasing lipolysis and, in general, affecting hepatic functions. Pro-inflammatory cytokine TNF-α was observed in serum of cows with fatty livers, and its concentration is well correlated with biomarkers of inflammation and the increment with a concomitant impairment of health and milk production [138]. It has been suggested that inflammatory-associated pathways are involved in adaptations to lactation. Many health disorders of dairy cows could be explained by the abrupt changes in nutrient requirements taking place in the lapse of calving, when the stressors lead the organism to oxidative stress. To mitigate the damage by oxidative stress through the maintenance of an adequate level of endogenous non-enzymatic antioxidants (BSA, glutathione) as well as enzymatic ones (superoxide dismutase (SOD), Catalase (CAT), glutathione peroxidase (GPx)) and exogenous antioxidant (vitamins, betaglucane, carotenoids, and flavonoids) according to the necessity, so it is important to know the requirements of the organism [139].

Transportation of animals is an essential component of livestock production and presents both economic and animal welfare concerns. Economic costs associated with transport are very high and negative data such as including morbidity, mortality, and carcass trim loss, and lower meat quality [39]. Transportation of cattle can contribute to the development of oxidative stress in several ways. Psychological stress and food deprivation stimulate fatty acid and amino acid mobilization for use as metabolic fuel, which increases mitochondrial ROS/ROSN production. During the transportation, the cows do not take food. This food deprivation and water deprivation provoke physical exertion in the animal that can stimulate an inflammatory response, which results in ROS/RONS production by phagocytic immune cells and as a by-product of eicosanoid biosynthesis. This costly condition, transportation stress, is a predisposing factor for bovine respiratory disease (BRD). In case of beef, animals may be transported several times during their lives: from birthplace to an auction market, stocker or backgrounding handling, feedlot, and finally to a processing facility implying great economic loss. Usually, when cattle have arrived to final destination, the privation of food and water continued until the animals are weighted. Public and consumers have increased their concern in how animals are raised and handled before slaughter, and the effects of transit on animal well-being is focus of interest [39].

As other organisms with highly developed central nervous systems, oxidative stress can be caused by psychological stressing factors derived from handling, fatigue, and potential for injury during the transport operation. The joined actions of different types of stressors activate several biological pathways known to lead to the development of oxidative stress. If some bioindicators of oxidative stress are compared between the calves that are not transported with those that are transported considerable distances, we find great differences between both groups. Lipid peroxidation was 218% greater, and antioxidant capacity was 186% lesser in leukocytes isolated from transported calves [140]. The same tendencies were found when cells were incubated with antioxidants (α-tocopherol or ascorbic acid) at concentrations greater than 0.1 mg/mL [141] as well as increased lipid peroxidation (177%) and decreased antioxidant status (11%) in the serum of beef calves after transit relative to pretransit values. The post-transit serum malondialdehyde (MDA) concentrations were 43% greater in calves, which later died of acute BRD, and calves that experienced ≥3 episodes of BRD had twofold greater MDA concentrations after transit than healthy calves [142]. It was observed greater linoleoyl tyrosine oxidation products before and after transit in blood samples from calves that eventually had pulmonary adhesions at slaughter, providing evidence that oxidative status may be a contributing factor in the development of BRD. ROS/RONS species produced by immune cells to eliminate pathogenic bacteria are able to damage and even kill the pulmonary cells, provoking inflammation in this damaged tissue resulting in respiratory dysfunction. These data indicate that transit-induced oxidative damage can impair immune-cell and respiratory function, which ultimately increases morbidity and mortality. Oxidative stress induced by transportation accomplished by increasing of respiratory rates because the effort to repair, degrade, or replace damaged biomolecules as well as deployment of antioxidant defense requires a great amount of energy. Indeed, a negative relationship between oxidative stress and feed efficiency in livestock has been observed and described [143]. During the trip, while trucks or cargo trains are moving, cows usually tend not to lie down, but when travel is very long, the animals will trait to lie down if adequate space is available.

5. General assessment to preventive strategies

Strategies based on dietary antioxidants may alleviate the impact of other sources of oxidative stress in farm animals and inhibit the negative influence of this stress on livestock production. These strategies generally involve reducing the concentration of polyunsaturated lipids in diets as well as supplementation with α-tocopherol (around 200 mg tocopherol/kg feed) and ascorbate (up to 1000 mg ascorbate/kg feed) alone or in combination with other elements with antioxidant potential, such as selenium, magnesium, and zinc in poultry [144–146]. The vitamin E, α-tocopherol, is an effective inhibitor of lipid peroxidation and well scavenger of free radicals such as peroxyl and alkoxyl and lipid radicals by transferring hydrogen, resulting in non-radical molecules. In addition to increasing the concentration of dietary antioxidants in tissues, these strategies, including dietary fat modification, may also support antioxidant protection by promoting the concentration and activity of endogenous antioxidant enzymes such as glutathione peroxidase (GSHPx). Selenium yeast may be, according to some recent reports, a promising dietary strategy to improve the oxidative stability in farm animals [90, 147]. In broiler chickens, dietary organic selenium improves antioxidant capacity and enhances growth performance in broiler chickens and the supplementation with algae-based selenium yeast in addition to improving not only the antioxidant defense capacity in live broilers, but also allowing to preserve the quality of the meat after slaughter by increasing lipid and protein oxidative stability of meat through promotion of antioxidant enzyme activity [148].

To establish the composition of antioxidant diet, it is necessary to define the specific nutritional requirements of the species, taking into consideration not only basic feeding parameters but also all the other feeding parameters would make essential micronutrients act as anti- or pro-oxidants. Molecular mechanisms triggered after feeding high-concentrate inducing sub-acute ruminal acidosis (SARA), a digestive disorder also known as chronic or subclinical acidosis, in dairy cows [149]. This is a relevant health problem in most dairy herds. This is digestive disorder characterized by changes in oxidative stress parameters, including the genomic signaling pathways in different organs and tissues: mammary gland, liver, hind-gut, and uterus and in epithelial tissue. Cows with SARA often develop complications or other diseases and associate physiologically with immunosuppression, inflammation, and oxidative stress. The prevention of SARA includes the establishment of feeding and management guidelines seeking to minimize rumen acidotic load including probiotics, fiber, and vitamins as well as yeast-derived supplements [150]. In our experience, this nutritional strategy improves overall animal performance in all aspects of their life and in productive parameters not only in cattle but also in other ruminants. A simple practical method for assessing the performances in dairy cows is the so-called body condition score (BCS). BCS helps to identify animals at increased risk of metabolic diseases and oxidative stress [151, 152]. By managing the BCS we can establish the relationships between animal health, reproductive function, milk production, and overall development of the cows. This index is associated with key hepatic enzymes associated with animal metabolism and related biomarkers, including oxidative stress biomarkers in liver tissue and plasma. Cows with high BCS show higher plasma concentrations of fatty acids compared with the other ones with normal BCS although concentration of reactive oxygen metabolites found in both groups was similar. High BCS cows showed lower overall concentrations of β-carotene and tocopherol, explaining the lower indicator (ferric reducing ability of plasma) of antioxidant capacity and showed lower hepatic protein abundance of the 1-carbon metabolism enzymes cystathionine-β-synthase, betaine-homocysteine methyltransferase, methionine adenosyltransferase 1 A, glutathione metabolism-related enzymes, and glutathione S-transferase α4 and GPx3. The published data suggest that the BCS values are well correlated with milk yield, immune response, and synthesis of antioxidant [126, 153, 154]. In deterioration of physical environment during the dry season and in the first time in body condition postpartum has been related to higher probabilities of metabolic and infection diseases [20, 155, 156]. The activity of specific additives is to efficiently manage the effect of oxidative stress derived from the determination of every specific functional element in the diet and determination of the optimized quantity and proportion in the feed formulation. Among such additives, we have amino acids, vitamins, microelements, prebiotics, and probiotics, and all of them must be continuously evaluated and valorized in animal nutrition, especially when natural antioxidant defenses have been breached. Little information exists on optimum inclusion levels and synergetic effects of rumen protected amino acids on their oxidative status. Increasing antioxidants activity has a beneficial effect on animal health and can decrease the incidence rate of metabolic disorder diseases such as ketosis [157]. Other additives are analogues that can be used to improve antioxidant defense of the organism like N-carbamylglutamate (NCG). NCG is a metabolically stable analogue of N-acetylglutamate synthase that produces endogenous arginine, and it is proven that can stimulate and improve the immune function and oxidative status in suckling lambs

[74]. Dietary supplementation with L-arginine (Arg) and N-carbamylglutamate (NCG) on intrauterine growth-retarded (IUGR) suckling lambs. Methionine supplementation coupled with choline enhanced gene expression of TLR2 and L-selectin, a part of pathogen recognition mechanisms [31]. Cells incubated without choline had high mRNA abundances encoding IL1B, IL6, IL10, myeloperoxidase (MPO), glutathi one reductase (GSR), GSS, cystathionine gamma-lyase (CTH), and cysteine sulfinic acid decarboxylase (CSAD), suggesting higher inflammation and oxidative stress [158, 159].

6. Practical examples

Our group carried out a series of experiments using a mixture of hepatoprotectors and antioxidant compounds (HPAC) in the stressed mice by intramuscular *via* for 1 month. HPAC mixture is an injectable aqueous solution composed of salts (sodium acetate, potassium chloride, calcium chloride, and magnesium sulfate), amino acids (phenylalanine, leucine, isoleucine, valine, tryptophan, arginine, cysteine, histidine, lysine, methionine, and threonine), vitamins (nicotinamide, pyridoxine, cyanocobalamin, riboflavin, pantenol, ascorbic acid, and tocopherol), and polysaccharides (dextrose, yeast β-glucan soluble fraction). Stressed mice were obtained according to the protocol described in the previous section and the same biomarkers were measured as well. The results showed that the mixture made up of hepatoprotectors and antioxidants at optimized proportions could prevent an oxidative stress-mediated damage in liver, kidney, and immunological functions. The obtained results clearly indicated the direct relationship between the excess of free radicals found in immune-depressed animals and their susceptibility to disease [159]. The results also showed interrelationship between important physiological functions and the oxidant/antioxidant balance in the organism. Non-treated control group presented alteration in oxidative status. In another research, a high-producer β-glucans strain of *S. cerevisiae* was

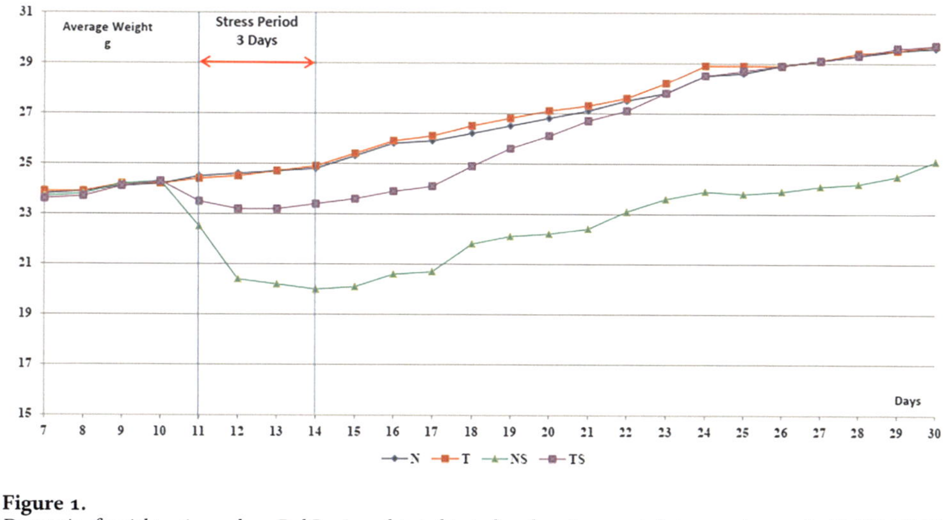

Figure 1.
Dynamic of weight gain week on BalC mice subjected to induced environmental stress and treated with HPAC. The treatments correspond to: N-not treated and not stressed (not stressed), NS-not treated and stressed, T-treated but not stressed, TS-treated and stressed. The best response to induced stress and the best weight recovery was observed in the group that received HPAC-treatment at days 1, 3, 6, and 9 of the experimental trial. The values shown in ***Tables*** *1 and 2 correspond to this experiment [10].*

selected from our culture collection to evaluate its ability to assimilate selenium by growing it in YPD (Yeast Extract, Peptone, Dextrose) medium supplemented with inorganic sodium selenite. This strain was also used as a host to express the murine lactoferrin gene under the control of the promoter of the *S. cerevisiae* glyceraldehyde-3-phosphate dehydrogenase (GPD) gene. The yeast strain was cultivated to obtain biomass made up of high β-glucans levels, the incorporated selenium and recombinant murine lactoferrin. This biomass was harvested and dried to obtain probiotic supplements T1 and T2. The amount of bioselenium and murine lactoferrin were determined in the resulting product and used to feed BALB/c mice during 30 days (**Figure 1**). Several parameters served to monitor evaluate the immune stimulatory effect and the physiological state of the animals during the test. Measurements were carried out at 0, 15th, and 30th days. The results showed the composite supplement improves the physiological and immunological conditions of the tested animals compared to the control group (**Tables 1** and **2**). The results obtained pave the way for developing food supplements with similar characteristics for economically important species [10].

Organic System	Test	T	Day 1	s	Day 15	s	Day 30	s	Observed Tendency
Hepatic	**Alanine Transaminase U/dL**	**ND**	26.41	1.57	26.45	2.74	27.93	2.15	Without significant changes
		NDT	25.37	1.56	42.5	1.65	54.76	198	Increment
		T	26.18	2.01	67.72	2.92	37.94	2.26	Increment, but recovering the normal values
	Aspartate Transaminase U/L	**ND**	65.38	1.95	66.41	2.05	67.42	2.58	Without significant changes
		NDT	63,14	1.65	86.76	1.43	89.83	1.74	Increment
		T	64.78	2.81	83.42	2.27	65.72	2.36	Increment, but recovering the normal values
	Alkaline Phosphatase U/L	**ND**	39.32	2.21	40.15	2.05	39.54	2.75	Without significant changes
		NDT	38.87	1.95	156.89	2.05	162.54	2.62	Increment
		T	40.07	2.32	142.95	2.15	56.56	1.59	Increment, but recovering the normal values
	Malondialdehide MDA Nmol/mg.pr	**ND**	3.56	1.47	3.51	1.48	3.53	1.54	Without significant changes
		NDT	3.49	1.23	7.05	1.36	12.05	1.04	Increment
		T	3.45	1.06	6.52	1.34	6.43	1.24	Increment, but recovering the normal values

Organic System	Test	T	Day 1	s	Day 15	s	Day 30	s	Observed Tendency
Renal	**Urea BUN mg/dL**	**ND**	20.41	1.45	21.01	1.32	21.32	1.43	Without significant changes
		NDT	22.21	1.23	20.43	1.32	22.23	1.35	Increment
		T	21.04	1.25	20.31	1.67	21.05	1.39	Without significant changes
	Uric Acid mg/dL	**ND**	0.15	0.01	0.16	0.01	0.15	0.01	Without significant changes
		NDT	0.12	0.03	0.16	0.04	0.12	0.02	Increment
		T	0.15	0.01	0.15	0.01	0.16	0.02	Without significant changes
	Creatinine mg/dL	**ND**	0.74	0.02	0.72	0.01	0.73	0.02	Without significant changes
		NDT	0.69	0.05	0.78	0.02	0.97	0.03	Increment
		T	0.75	0.03	0.71	0.02	0.73	0.03	Without significant changes
Serum Oxidative Status	**Serum Albumin mg/dl**	**ND**	2.61	0.01	2.65	0.12	2.67	0.03	Without significant changes
		NDT	2.56	0.02	3.04	0.11	3.20	0.19	Increment
		T	2.73	0.01	3,15	0.12	2.69	0.03	Increment, but recovering the normal values
	Serum Glutathione Peroxidase * (GSH-Px) nmol/ml	**ND**	28.5	1.09	28.18	1.34	29.31	1.81	Without significant changes
		NDT	26.99	1.43	29.87	1.28	48.65	1.58	Increment
		T	27.7	1.38	28.73	1.68	28.89	1.21	Increment, but recovering the normal values
	Serum Total Antioxidant Capacity nmol/L	**ND**	0.96	0.04	0.92	0.04	0.96	0.05	Without significant changes
		NDT	0.95	0.06	1.89	0.07	1.99	0.08	Increment
		T	0.97	0.04	1.42	0.06	1.08	0.06	Increment, but recovering the normal values

*ND: No supplemented diet no stressed; NDT: no supplemented diet stressed; T: Supplemented diet stressed. Biomarkers in hepatic, renal and serum oxidative status show some increments in individuals fed with a supplemented diet, but at the level of the entire organism no retard was observed in their development, weight gain or behavior; compared with individuals that were fed with normal diet (**Figure 1**) [10].*

Table 1.
Blood biochemical tests performed in peripheral blood of BALB/c mice to evaluate hepatic, renal and serum oxidative status.

Biomarker	Test	T	Day 1	s	Day 15	s	Day 30	s	Observed Tendency
Leucocytes Cells x 10^3/ μL	**Lymphocytes Cells x 10^3/μL**	**ND**	12.23	0.43	12.55	0.73	12.29	1.05	Without significant changes
		NDT	12.54	0.65	16.47	0.64	18.65	0.54	Increment
		T	12.08	0.71	16.31	0.87	18.55	0.32	Increment
	Neutrophils Cells x 10^3/μL	**N**	6.51	0.41	6.71	0.51	6.23	1.15	Without significant changes
		NDT	6.09	0.32	7.89	0.43	7.45	1.20	Increment
		T	6.88	0.52	7.84	0.61	8.86	1.07	Increment
	Monocytes Cells x 10^3/μL	**ND**	0.74	0.23	0.79	0.12	0.79	0.15	Without significant changes
		NDT	0.69	0.13	0.81	0.12	0.91	0.12	Increment
		T	0.79	0.16	0.89	0.16	0.91	0.14	Increment
Phagocytosis assay (%)	**Phagocytosis in Monocytes (%)**	**ND**	36.09	1.52	38.55	1.29	37.49	1.32	Without significant changes
		NDT	35.12	1.45	37.09	1.61	65.05	1.28	Increment
		T	35.68	1.71	57.21	1.53	70.63	1.51	Increment
	Phagocytosis in Macrophages (%)	**N**	33.21	1.31	35.57	1.45	35.23	1.51	Without significant changes
		NDT	33.42	1.03	57.5	1.26	66.98	1.34	Increment
		T	35.88	1.22	67.34	1.05	78.16	1.42	Increment
Cytokines pg/L	**Interferon gamma (IFN-γ). pg/L**	**ND**	130.19	1.27	141.21	1.15	149.25	1.54	Without significant changes
		NDT	139.04	1.54	1.49.97	1.24	216.34	1.33	Increment
		T	141.31	1.79	187.95	1.38	312.55	1.67	Increment
	Interlukin 2 (IL2) pg/L	**N**	210.19	1.32	200.52	0.13	231.49	0.59	Without significant changes
		NDT	216.45	1.27	234.09	0.18	356.43	0.76	Increment
		T	221.31	0.79	289.92	1.62	382.55	0.63	Increment
	Interlukin 12 (IL2) pg/L	**N**	322.45	1.05	327.31	1.61	339.67	1.25	Without significant changes
		NDT	310.45	1.25	315.32	1.57	456.87	1.49	Increment
		T	315.86	1.32	436.57	1.47	532.48	1.37	Increment
	Interlukin 4 (IL2) Pg/L	**N**	273.81	1.12	296.45	1.31	326.68	1.09	Without significant changes

Biomarker	Test	T	Day 1	s	Day 15	s	Day 30	s	Observed Tendency
		NDT	267.98	1.25	338.76	1.47	367.78	1.26	Increment
		T	281.23	1.28	351.2	1.39	425.82	1.46	Increment
	Interlukin 10 (IL2) pg/L	N	359.51	1.41	373.65	1.46	321.43	0.75	Without significant changes
		NDT	361.09	1.45	389.65	1.34	423.67	1.25	Increment
		T	348.23	.1.49	537.69	1.51	754.31	0.65	Increment

*ND: No supplemented diet no stressed; NDT: no supplemented diet stressed; T: Supplemented diet stressed. The treated individuals are in a state of alert that does not represent damage to the vital functions but shows that increments in immune functions are observed in both individuals fed with supplemented diet and stressed as well as in individuals fed with no supplemented diet and stressed the values in immune functions were lower than in those fed with supplemented diet (**Figure 1**) [10].*

Table 2.
Analysis of immunological status leucocyte blood cells counts and cytokine production in BALB/c mice.

7. Concluding remarks

The role of oxidative stress on biological systems is important but controversial, it is important in organic functions particularly in signal transduction. Oxygen participates in an important way in the energy cycle of living beings. It is essential for cellular respiration in aerobic organisms but many of living process generate reactive oxygen species and reactive oxygen and nitrogen species (ROS/RONS) that in excess can drive the organism to oxidative stress, a consequence of the disturbance of the balance between pro-oxidants and antioxidants, with a shift in favor of the ROS/RONS and significantly damaging the physiological functions of the organism. The aerobic organisms should keep the oxidant/antioxidant balance using different antioxidant mechanisms. The cellular antioxidant defenses include the use of free amino acids, vitamins, proteins, polysaccharides, sugars, and certain ions as antioxidant compounds as well as enzymes such as superoxide dismutase and (SOD), glutathione peroxidase, and catalases.

A group of stressors contributes to a greater generation of ROS/RONS that can be caused by the intracellular and organic metabolic processes themselves, and factors of infection by pathogens and environmental conditions. Oxidative stress is a permanent research target due to the importance it generates for human health, and many investigations materialize in different study models from bacteria, simple eukaryotes such as yeast, cell lines, first culture cells, animals, and plants. Using accepted models that exist for different diseases, we can see how the oxidative condition of the organism influences each of these diseases. Oxidative stress also affects the condition of all living organisms and significantly affects all living organisms, including those that are of economic interest, particularly under the conditions imposed by intensive animal production, so the stress is generates due to overpopulation in farms and management in the different steps of the production cycle. The incidence of oxidative stress in animals, both experimental models and farm production animals, was analyzed in some species of interest. To prevent and improve the damage caused by the digestive process, the use of active antioxidants in food is recommended, as well as improvements in living conditions.

Author details

Gustavo Alberto De La Riva De La Riva[1*], Luis Adrián Saldaña Trujillo[1]
and Juan Carlos González-Hernández[2]

1 Departamento de Bioquímica, Instituto Tecnológico de la Piedad, Tecnológico Nacional de México, La Piedad, Michoacán, México

2 Instituto Tecnológico de Morelia, Tecnológico Nacional de México, Morelia, Michoacán, México

*Address all correspondence to: griva_2010@hotmail.com;
uustavo.rr@piedad.tecnm.mx

References

[1] Lyons TW, Reinhard CT, Planavsky NJ. The rise of oxygen in Earth's early ocean and atmosphere. Nature. 2014;**506**(7488):307-315

[2] Halliwell B. Free radicals and antioxidants: Updating a personal view. Nutrition. Reviews. 2012; **2577**(5):265

[3] Durand D, Damon M, Gobert M. Le stress oxydant chez les animux de rente: principes généraux. Nutrition Animale. 2012;**48**:218-224

[4] Massaad CA, Klann E. Reactive oxygen species in the regulation of synaptic plasticity andmemory. Antioxidants and Redox Signaling. 2011; **14**(10):2013-2054

[5] Bild W, Ciobica M, Padurariu M, Bild V. The interdependence of the reactive species of oxygen, nitrogen, and carbon. Journal of Physiology and Biochemistry. 2013;**69**(1):147-154

[6] Kumar B, Manuja A, Aich P. Stress and its impact on farm animal. Frontiers in Bioscience. 2012;**E4**:1759-1767

[7] Lykkesfeldt J, Svendsen O. Oxidants and antioxidants in disease: Oxidative stress in farm animals. The Veterinary Journal. 2007;**173**:502-511

[8] Sordillo LM, Mavaangira V. The nexus betwen nutrient metabolism, oxidative stress and inflamation in transition cows. Animal Production Science. 2014;**54**:1204-1214

[9] De la Riva GA, Lopez-Mendoza FJ, Agüero-Chapin G. Known Hepatoprotectors act as antioxidants and immune stimulators in stressed mice. Perspectives in animal health care. Current Pharmaceutical Design. 2018; **24**(40):4825-4837. DOI: 0.2174/13816128256

[10] De la Riva G, Torres-Osornio A, Agüero-Chapín G, Saldaña-Trujillo L, Arellano-Nuñez A, López-Mendoza F. Promising probiotic food supplement based on combination of yeast beta-glucan, bioselenium and lactoferrin for animal health. Biomedical Journal of Scientific Technical Research. 2020;**32**(4): 25105-25113

[11] Daiber A, DiLisa F, Oelze M, Kroller-Schon S, Steven S, Munzel T. Crosstalk of mitochondria with NADPH oxidase via reactive oxygen and nitrogenspecies signalling and its role for vascular function. British Journal of Pharmacology. 2015;**74**:1670-1689

[12] Dunnill C, Patton T, Brennan J, Barrett J, Dryden M, Cooke J, et al. Reactive oxygen species (ROS) an wound healing: The functional role of ROS and emerging ROS-modulatinng technologies for augmentation of the healing process. International Wound Journal. 2017;**14**:89-96

[13] Friedman JR, Nunnari J. Mitochondrial form and function. Nature. 2014;**505**:335-343

[14] Schiavone S, Jaquet V, Trabace L, Krause KH. Severe life stress and oxidaative stress in the brin: From animal model to human pathoogy. Antioxidants Redox Signaling. 2013;**18**(12):1475-1490

[15] Eugene G, Hrycay EG, Bandiera SM. Involvemment of Cytocrome P450 in reactive oxygen species formation and cancer. Advances in Pharmacology. 2015;**44**:35-84

[16] Fransen M, Nordgren M, Wang B, Apanas O. Role of peroxisomes in

ROS/RNS-metabolism: Implications for human disease. Biochimica et Biophysica Acta (BBA) - Molecular Basis of Disease. 2012;**1822**(9):1363-1373

[17] Dupré-Crochet S, Erard M, Nü O. ROS production in phagocytes: Why, when and where? Journal of Lukocyte Biology. 2013;**94**:657-670

[18] Sekine S, Ito K, Watanabe H, Nakano T, Moriya K, Shintani Y, et al. Mitochondrial iron accumulation exacerbates hepatic toxicity caused by hepatitis C virus core protein. Toxicology and Applied Pharmacology. 2015;**282**(3):237-243

[19] Halliwell B. Free radicals and antioxidants—Quo vadis? Trends in Pharmacological Sciences. 2011;**32**(3): 125-130

[20] Cichoz-Lach H, Michlak A. Oxidative stress as a crucial factor in liver diseases. World Journal of Gastroenterology. 2014;**20**(25): 8082-8091. DOI: 10.3748/wjg.v20

[21] Egea J, Fabregat I, Frapart Y, Ghezzi P, Görlach A, Kietzmann T, et al. European contribution to the study of ROS: A summary of the findings and prospects for the future from the COST action BM1203 (EU-ROS). Redox Biology. 2017;**13**:94-162

[22] Samet JM, Wages PA. Oxidative stress from environment exposures. Current Opinion in Toxicology. 2018;**7**:60-66

[23] Suwanprinya L, Phumala N, Sanvarinda P, Dieng H, Okabayashi T, Morales-Vargas RE. Reactive oxygen species production in rat microglial cells. Japanese Journal of Infection. 2017; **70**(4):383-387

[24] Zahlten J, Kim Y, Doehn J, Pribyl TH, García P, Hammerschmidt S, et al. Streptococcus pneumoniae-induced oxidative stress in lung epithelial cells on pneumococcal autolysis and is reversible by resveratrol. The Journal of Infectious Diseases. 2015;**211**(11):1822-1830

[25] Baruch M, Baruch B, Ravins M, Anand A, Cheng C, Biswas D, et al. Induction of endoplasmatic reticulum stress and unfolded protein response constitutes a pathogenic strategy of group a streptococcus. Frontiers in Cellular and Infection Microbiology. 2014;**4**:105

[26] Cheresh P, Kim SJ, Tulasiran S, Kamp DW. Oxidative stress and pulmonary fibrosis. Biochem Biophys Acta. 2013;**1832**:1028-1040

[27] O'Neill S, Brault J, Stasia MJ, Knaus UG. Genetic disorders coupled to ROS deficiency. Redox Biology. 2015;**6**: 135-156

[28] Cabiscol E, Ros J. Oxidative damage to proteins: Structural modifications and consequences in cell function. In: Dale-Dome I, Scaloni A, Butterfield DA, editors. REDOX Proteomics. New Jersey: Wiley Interscience; 2006. pp. 247-281

[29] Bedard K, Krause K. The NOX family of ROS generating HAADPH oxydases: Physiology and pathophysiology. Physiological Reviews. 2007;**192**:1001-1114

[30] Rojas-Rivera C, Hetz TM. BIM protein family: Ancestral regulators of cell death. Apoptosis. 2009;**14**:1255-1265

[31] Hu W, Spaink HP. The role of TLR2 in infectious diseases caused by mycobacteria: From cell biology to therapeutic target. Biology. 2022;**11**(2):246

[32] Klotz LO, Sanchez-Ramos C, Prieto-Arroyo I, Urbanek P, Steinbrenner H, Monsalve M. Redox regulation of

FoxO1a transcription factors. Redox Biology. 2015;**6**:51-72

[33] Mikhed Y, Görlach A, Knaus UG, Daiber A. A redox regulation of genome stability effects on gene expression, epigenetic pathways and DN damage/ repair. Redox Biology. 2015;**5**:275-289

[34] Baird AM, O'Byrne K, Gray S. Reactive oxygen species and reactive nitrogen species in epigenetic modifications. In: Laher I, editor. Systems Biology of Free Radicals and Antioxidants. Heidelberg: Springer Berlin Heidelberg; 2014. pp. 437-455

[35] Lobo V, Patil A, Phatak A, Chandra N. Free radicals, antioxidants and functional food: Impact on human health. Pharmacological Reviews. 2010; **4**(8):118-126

[36] Tarafdar A, Pula G. The role of NADPH oxidases and oxidative stress in neurodegenerative disorders. International Journal of Molecular Sciences. 2018;**19**(12):3824

[37] Isaksson C. Urbanization, oxidative stress and inflammation: A question of evolving, acclimatization or coping with urban environmental stress. Functional Ecology. 2015;**29**:913-923

[38] Habeeb AA. Oxidative stress in animals exposed to different stressful conditions. International Journal of Nutritional Science. 2018;**3**(2):1027

[39] Deters EL, Hansen L. Linking road transportation with oxidative stress in cattle and other species. Applied Animal Science. 2020;**36**(2):183-200

[40] Imlay J. The molecular mechanisms and physiological consequences of oxidative stress: Lessons from a model bacterium. Nature Reviews. Microbiology. 2013;**11**:443-454

[41] Mih N, Monk J, Fang X, Catoiu E, Heckmann D, Yang L, et al. Adaptations of *Escherichia coli* strains to oxidative stress are reflected in properties of their structural proteomes. BMC Bioinformatics. 2020;**21**(1):162

[42] Imlay JA. Where in the world do bacteria experience oxidative stress? Environmental Microbiology. 2019;**21**: 521-530

[43] Forsburg S. The art and design of genetic screens: Yeast. Nature and Reviews Genetics. 2001;**2**:659-668

[44] Scherens B, Goffeau A. The uses of genome-wide mutant collections. Genome Biology. 2004;**5**:229

[45] Picazo C, Molin M. Impact of hydrogen peroxide on protein synthesis in yeast. Antioxidants. 2021;**10**(6):952

[46] Bhvya G, Hiremath KY, Jogaiah S, Geetha N. Heavy metal-induced oxidative stress and alteration in secretory proteins in yeast isolates. Archives of Microbiology. 2022;**204**(3): 1-11

[47] Fasnacht M, Polacek N. Oxidative stress in bacteria and the central dogma of molecular biology. Frontiers in Molecular Biosciences. 2021;**8**:392

[48] Lee Y, Reader B, Dumont C, Lee Y, Mokadam N, Black S, et al. Pegylated-catalase is protective in lung ischemic injury and oxidative stress. The Annals of Thoracic Surgery. 2021;**111**(3): 1019-1027

[49] Gualtieri R, Kalthur G, Brbato G, Di Nardo M, Adiga SK, Talevi R. Mitochondrial dysfunction and oxidative stress caused by cryopreservation in reproductive cells. Antioxidants. 2021; **10**(3):337

[50] Duan J, Gao S, Tu S, Lenahan C, Shao A, Sheng J. Pathophysiology and therapeutic potential of NADPH oxidases in ischemic stroke-induced oxidative stress. Oxidative Medicine and Cellular Longevity. 2021; **2021**:1-11

[51] Halliwell B. Cell culture, oxidative stress, and antioxidants: Avoiding pitfalls. Journal of Biomedical Science. 2014;**37**(3):99-105

[52] Schwartz M, Kipnis J, Rivest S, Prat A. How do immune cells support and shape the brain in health, disease and aging. Neuroscience. 2013;**33**: 17587-175996

[53] Zhang H, Huang L, Wang J, Gong L, Liu J, Sun B. Evaluation of the cellular and animal models for the study of antioxidant activity: A review. Journal of Food Science. 2017;**82**:1-11

[54] Wang CY, Liu LN, Zhao ZB. The role of ROS toxicity in spontaneous aneuploidy in cultured cells. Tissue & Cell. 2013;**45**:47-53

[55] Wan N, Xu Z, Liu T, Min Y, Li S. Cadmium-induced apoptosis through the mitochondrial pathway in rainbow trout hepatocytes: Involvement of oxidative stress. Aquatic Toxicology. 2018;**69**:247-258

[56] Klimova Y, Chuikoa G, Pesnyaa D, Ivanova E. Biomarkers of oxidative stress in freshwater bivalve mollusks. Inland Water Biology. 2020;**13**(4): 674-683

[57] Timothy J, Kraev I, Lange S. Extracellular vesicles and post-translational protein mussel (*Mytilus edulis*), soft Shell clam (*Mya arenaria*) eastern oyster (*Crassostrea virginica*) and Atlantic Jacknife clam (*Ensis leei*). Biology. 2020;**9**:416

[58] López-Pedrouso L, Lorenzo JM, Varela Z, Fernández JA, Franco D. Finding biomarkers in antioxidant molecular mechanisms for ensuring food safety of bivalves threatened by marine pollution. Antioxidants. 2022;**11**:369

[59] Manduzio H, Rocher B, Durand F, Galap C, Leboulnger F. The point about oxidative stress in molluscs. Invertebrate Survival Journal. 2005;**2**:91-104

[60] Soldatov AA, Gostyukhina OL, Golovina IV. Antioxidant enzyme complex of tissues of the bivalve *Mytilus galloprovincialis* lam. Under normal and oxidative-stress conditions: A review. Applied Biochemistry and Microbiology. 2007;**43**:556-562

[61] Primost M, Sabatini S, Di Salvatore P, Rios De Molina M, Bigatti G. Oxidative stress indicators in populations of the gastropod *Buccinanops globulosus* affected by imposex. Journal of the Marine Biological Association of the United Kingdom. 2017;**97**(1):35-42

[62] Canesi L, Caterina C, Fabbri R, Marcomini A, Pojana G, Gallo G. Bivalve molluscs as a unique target group for nanoparticle toxicity. Marine Environmental Research. 2012;**76**:16-21

[63] Krishnakumar PK, Casillas E, Snider RG, Kagley AN. Environmental contaminants and the prevalence of hemic neoplasia (leukemia) in the common mussel (*Mytilus edulis*) from Puget Sound, Washington, USA. Journal of Invertebrate Pathology. 1999;**73**: 135-146

[64] Kournoutou G, Giannopoulou P, Sazaki H, Leotsinidis M, Kalpaxis D, Dinos G. Oxidative damage of mussels living in seawater enriched with trace metals, from the viewpoint of proteins expression and modification. Toxics. 2020;**8**:89

[65] Van de Braak CB, Botterblom MH, Taverne N, Muiswinkel WB, Rombout Van der Knaap WP. The roles of haemoytes and the lymphoid organ in the clearance of injected vibrio bacteria in *Penaeus monodon* shrimp. Fish & Shellfish Immunology. 2002;**13**(4): 293-309

[66] Venero Gutiérrez JR. Oxidative damage, free radicals and antioxidants. Rev. Cub. Med. Mil. 2002;**31**(2):126-133

[67] Hartinah T, La Paturusi LP, Ratsanari R, Rimal H, Dahlia D, Rustam R. Performance of total haemocyte count and survaival rate the tiger *Penaeus monodon fabricius* juvenile rearing at high density. Aquaculture Indonesia. 2017;**18**:9-14

[68] Benítez-Santana T, Masuda R, Juárez-Carrillo E, Ganuzza E, Valencia A, Hernández-Cruz C, et al. Dietary n-3 HUFA deficiency induces a reduced response in gilthead seabream *Sparus aurata* larvae. Aquaculture. 2007; **264**:408-417

[69] An W, He H, Dong X, Tan B, Yang Q, Chi S, et al. Regulation of groeth, fatty acid profiles, hematological characteristic and hepatopancreatic histology by different dietary n-3 highly unsaturated fatty acuds levels in the first stags of juvenile Pacific white shrimp (*Litopenaus vannamei*). Aquaculture Reports. 2020;**17**:100321

[70] Ma J, Shao Q, Xu Z, Zhou F. Effect of dietary n-3 highly unsaturated fatty acids on growth, body composition and fatty acid profiles of juvenile black seabram *Acanthopargus schlegeli* (Bleeker). Journal Would Aquaculture Society. 2013;**44**:311-325

[71] Jin J, Blckwood E, Azizi K, Thuerauf D, Fahem A, Hofmann C, et al. ATF6 decreases myocardial ischemia/reperfusion damage and links ER stress and oxidative stress signaling pathways in the heart. Circulation Research. 2017; **120**:862-875

[72] Ke C, Erchao L, Chang X, Xiaodan W, Huifeng L, Jian GQ. Growth and metabolic response of Pacific white shrimp (*Litopenaeus vannamei*) to different dietary fatty acid sources and salinity levels. Aquaculture. 2019;**499**: 329-340

[73] González-Felix ML, Lawrence AL, Gatlin DM, Pérez-Velázquez M. Growth, survival and fatty acid composition of juvenile *Litopenaeus vannamei* fed different oils in the presence and absence of phopholipids. Aquaculture. 2002;**235**: 325-343

[74] Zhang H, Zhao F, Peng A, Dong L, Wang LY, et al. Effects of dietary l-arginine and N-carbamylglutamate supplementation on intestinal integrity, immune function, and oxidative status in intrauterine-growth-retarded suckling lambs. Journal of Agriculture and Food Chemistry. 2018; **66**(16):4145-4154

[75] Zainuddin H, Haryati H, Aslmyah S. Effect of dietary carbohydrate levels and feeding frequencies on growth and carbohydrate digetibility of white shrimp *Litopenaeus vannamei* under laboratory conditions. Journal of Aquaculture Research and Development. 2014;**5**:274

[76] Ayisi CL, Hua X, Apraku A, Afriyie A, Amankwah KB. Recent studies toward the development of practical diets for shrimp and their nutritional requirements. Journal of Biosciences. 2017;**24**:109-117

[77] Ruvalcaba-Márquez J, Álvarez-Ruíz P, Zenteno-Savín T, Martínez Antonio E, Goytortúa-Bores E, Casillas-Hernández R,

et al. Performance, immune response, and oxidative stress parameters of *Litopenaeus vannamei* fed diets containing varying carbohydrate/protein, lipid/protein, and energy/protein ratios. Aquaculture Reports. 2021;**21**:100771

[78] Mehjbeen J, Irshad A, Masood A. Bioaccumulation, oxidative stress and genotoxicity in fish (*Channa punctatus*) exposed to a thermal power plant effluent. Ecotoxicology and Environmental Safety. 2016;**127**:163-169

[79] Qiu M, Zuo Q, Wu Q, Yang Z. Water ecological security assessment and spatial autocorrelation analysis of prefectural regions involved in the Yellow River Basin. Scientific Reports. 2022;**12**:5105

[80] Abbate F, Maugeri A, Laurà R, Levanti M, Navarra M, Cirmi S, et al. Zebrafish as a useful model to study oxidative stress-linked disorders: Focus on flavonoids. Antioxidants. 2021;**10**:668

[81] Chen Y, Yang Z, Wen C, Chang Y, Wang B, Hsiao C, et al. Evaluation of the structure-activity relationship of flavonoids as antioxidants and toxicants of zebra fish larvae. Food Chemistry. 2012;**134**:717-724

[82] Cirmi S, Ferlazzo N, Lombardo G, Ventura-Spagnolo E, Gangemi S, Calapai G, et al. Neurodegenerative diseases: Might citrus flavonoids play a protective role? Molecules. 2016;**21**:1312

[83] Ferlazzo N, Cirmi S, Calapai G, Ventura-Spagnolo E, Gangemi S, Navarra M. Anti-inflammatory activity of *Citrus bergamia* derivatives: Where do we stand? Molecules. 2016;**21**:1273

[84] Mannucci C, Casciaro M, Sorbara E, Calapai F, Di Salvo E, Pioggia G, et al. Nutraceuticals against oxidative stress in autoimmune disorders. Antioxidants. 2021;**10**:261

[85] Belton B, Thilsted SH. Fisheries in transition: Food and nutrition security implications for the global south. Global Food Security. 2014;**1**:56-66

[86] Béné C, Barange M, Subasinghe R, Pinstrup-Andersen P, Merino G, Gro-Ingunn H, et al. Feeding 9 billion by 2050 - putting fish back on the menu. Food Security. 2015;**7**:261-264

[87] FAO. Sustainability in action. In: The State of World Fisheries and Aquaculture 2020. Rome: The State of World Fisheries and Aquaculture (SOFIA); 2020

[88] Schreck C, Contreras-Sanchez W, Fitzpatrick M. Effects of stress on fish reproduction, gamete quality, and progeny. Aquaculture. 2001;**197**:3-24

[89] Sadovy Y. The threat of fishing to highly fecund fishes. Journal of Fish Biology. 2001;**59**:90-108. DOI: 10.1111/j.1095-8649.2001.tb01381.x

[90] Surai PF, Kochish II, Fisiniin VI, Kidd MT. Antioxidant defense systems and oxidative stress in poultry biology: An update. Antioxidants. 2019;**8**:235

[91] Zhang J, Li C, Tng X, Lu Q, Sa R, Zhang H. High concentrations of atmospheric ammonia induce alterations in the hepatic proteome of broilers (*Gallus gallus*): An iTRAQ-based quantitative proteomic analysis. PLoS One. 2015;**10**:e0123596

[92] Sihvo HK, Immonen K, Poulanne E. Myodegeneration with fibrosis and regeneration in the pectoralis major muscle of broilers. Veterinary Pathology. 2013;**51**:619-623

[93] Yildiz D. Global poultry industry and trends. Feed Additive. 2022;**16**:1-4

[94] Estévez M. Oxidative damage to poultry: From farm to fork. Poultry Science. 2015;**94**(6):1368-1378

[95] Mishra B, Jha R. Antioxidants: Their effects on broiler oxidative stress and its meat oxidative stability. Frontiers in Veterinary Science. 2019;**6**:60

[96] Flees J, Rajaei-Sharifabadi H, Greene E, Beer L, Hargis B, Ellestad L, et al. Effect of Morinda citrifolia (noni) enriched diet on hepatic heat shock protein and lipid metabolism-related genes in heat stressed broiler chickens. Frontiers in Physiology. 2017;**8**:919

[97] Laliotis GP, Bizelis I, Rogdaki E. Comparative approach of the de novo fatty acid synthesis (lipogenesis) between ruminant and non ruminant mammalian species: From biochemical level to the main regulatory lipogenic genes. Current Genomics. 2010;**11**: 168-183

[98] Surai P, Fisinin VI. Selenium in poultry breeder nutrition: An update. Animal Feed Science and Technology. 2014;**191**:1-15. DOI: 10.1016/j.anifeedsci.2014.02.005

[99] Iqbal M, Pumford N, Tang Z, Lassiter K, Ojano-Dirain C, Wing T, et al. Compromised liver mitochondrial function and complex activity in low feed efficient broilers are associated with higher oxidative stress and differential protein expression. Poultry Science. 2005;**84**(6):933-941

[100] Debski B, Zalewski W, Gralak MA, Kosla T. Cr-yeast supplementation of chicken broilers in an industrial farming system. Journal of Trace Elements in Medicine and Biology. 2004;**18**:47-51

[101] Jackson A, Powell S, Johnston S, Shelton J, Bidner T, Valdez F, et al. The effect of Cr propionate on growth performance and carcass traits in broilers. Journal of Applied Poultry Research. 2008;**17**:476-481

[102] Zikov Balos M, Ljubojevic D, Jaksic S. The role and importance of vanadium, chromium and nickel in poultry diet. World's Poultry Science Journal. Mar 2017;**73**(1):5-16

[103] Bangyuan W, Hengmin C, Xi P, Jing F, Zhicai Z, Junliang D, et al. Investigation of the serum oxidative stress in broilers fed on diets supplemented with Ni chloride. Health. 2013;**5**:454-459

[104] Rayman MP. Selenium and human health. Lancet. 2012;**379**(156):68

[105] Roman M, Jitaru P, Barbante C. Selenium biochemistry and its rol for human health. Metallomics. 2014;**6**(1): 25-54

[106] Habibian M, Sadeghi G, Ghazi S, Mehdi Moeini M. Selenium as a feed Suplement for heat-stressed poultry: A review. Biological Trace Element Research. 2015;**165**:183-193

[107] Faverney C, Orsini N, de Sousa G. Cadmium-induced apoptosis through the mitochondrial pathway in rainbow trout hepatocytes: Involvement of oxidative stress. Aquatic Toxicology. 2004;**69**:247-258

[108] Julin J, Murphy B, Patoulias D, Fountoukis C, Olenius T, Pandis S, et al. Impacts of future European emission reductions on aerosol particle number concentrations accounting for effects of ammonia, amines, and organic species. Environmental Science Technology. 2018;**52**(2):692-700

[109] Jing H, Wang S, Shen N, Gao X. Environmental contaminant ammonia triggers epithelial-to-mesenchymal

transition-mediated jejunal fibrosis with the disassembly of epithelial cell-cell contacts in chicken. Science of the Total Environment. 2020;**726**:138686

[110] Arjumand R, Mohd T, Sheik A, Azmat A, Saim Q, Madeeha U, et al. Effect of chemically treated litter on ammonia Emmission, performance and carcass characteristics of broiler chicken. Journal of World Poultry Research. 2017; **7**(2):88-93

[111] Gibbs R, Weinstock G, Metzker M, Sodergren D, Scherer S, Scott G, et al. Rat genome sequence project consortium. Genome sequence of the Brown Norway rat yields insights into mammalian evolution. Nature. 2004; **428**(6982):493-521

[112] Graur D, Duret L, Gouy M. Phylogenetic position of the order Lagomorpha (rabbits, hares and allies). Nature. 1996;**379**:333-335

[113] Soussi T, Wiman KG. Shaping genetic alteration in human caner: The p53 mutation paradigm. Cancer Cell. 2007;**12**(4):303-312

[114] Levatic J, Salvadores M, Fuster-Torno F, Supek F. Mutational signatures are markers of drug sensitivity of cancer cells. Nature Communications. 2022; **13**(1):1-19

[115] Pouyet L, Carrier A. Mutant mouse models of oxidative stress. Transgenic Research. 2010;**19**:155-164

[116] Hamilton RT, Walsh ME, Van Remmen H. Mouse models of oxidative stress indicate a role for modulating healthy aging. Journal of Clinical Experimental Pathology. 2012;**4**: 005

[117] Oikawa D, Akai R, Tokuda M, Takao IT. A transgenic mouse model for monitoring oxidative stress. Scientific Reports. 2012;**2**:229

[118] Yildiz D. Global pig industry and trends. Feed Additives. 2021;**10**:72-78

[119] The Livestock Conservancy Annual Report 2016-2017, (2016). Conservation Priority List. Pitsboro, NC: The Livestock Conservation

[120] Campbell JM, Crenshaw JD, Polo J. The biological stress of early weaned piglets. Journal of Animal Science and Biotechnology. 2013;**4**(1):19

[121] Spreeuwenberg NA, Verdonk JM, Gaskins HR, Verstegen MW. Small intestine epithelial barrier function is compromised in pigs with low feed intake at weaning. The Journal of Nutrition. 2001;**131**:1520-1527

[122] Wang H, Zeng X, Zhang X, Liu H, Xing H. Ammonia exposure induces oxidative stress and inflammation by destroying the microtubule structures and the balance of solute carriers in the trachea of pigs. Ecotoxicology and Environmental Safety. 2021;**212**:111974

[123] Grzela K, Litwiniuk M, Zagorska W, et al. Airway remodeling in chronic obstructive pulmonary disease and asthma: The role of matrix Metalloproteinase-9. Archivum Immunologiae et Therapiae Experimentalis. 2016;**64**:47-55

[124] Rosas-Valencia U, Ortega-Cerrilla ME, Pérez Rodrigues P, Ayala-Rodriguez JM, Aranda Osorio G, Sanchez Torres-Esqueda MT. Beef cattle production with animal welfare. Agroproductividad. 2019;**12**(10):41-46

[125] Ayemele A, Tilahun M, Lingling S, Elsaadawy S, Guo Z, Zhao G, et al. Oxidative stress in dairy cows: Insights into the mechanistic mode of actions and

mitigating strategies. Antioxidants. 2021; **10**(12):1918

[126] Xue M, Sun H, Wu X, Guan LL, Liu J. Assessment of rumen microbiota from a large dairy cattle cohort reveals the pan and Core Bacteriomes contributing to varied phenotypes. Applied and Environmental Microbiology. 2018;**8419**:e00970-e00918

[127] Sørensen L, Mark T, Madsen P, Lund M. Genetic correlations between pathogen-specific mastitis and somatic cell count in Danish Holsteins. Journal of Dairy Science. 2009;**927**:3457-3471

[128] Tong J, Zhang H, Zhang Y, Xiong B, Jiang L. Microbiome and metabolome analyses of Milk from dairy cows with subclinical *Streptococcus agalactiae* mastitis—Potential biomarkers. Frontiers in Microbiology. 2019;**10**:547

[129] Bicalho ML, Machado VS, Higgins CH, Lima FS, Bicalho RC. Genetic and functional analysis of the bovine uterine microbiota. Part I: Metritis versus healthy cows. Genetic and functional analysis. 2017;**100**(5): 3850-3862

[130] Kononoff PJ, Heinrichs AJ. Evaluating Particle Size of Forages and TMRs Using the New Penn State Forage Particle Separator; a, 2002. State College, PA, US: Penn State University; 2002

[131] Jeon S, Lima F, Vieira-Neto A, Machado V, Lima S, Bicalho R, et al. Shift of uterine microbiota associated with antibiotic treatment and cure of metritis in dairy cows. Veterinary Microbiology. 2018;**214**:132-139

[132] Humer E, Petri RM, Aschenbach JR, Bradford BJ, Penner GB, Tafaj M, et al. Invited review: Practical feeding management recommendations to mitigate the risk of subacute ruminal acidosis in dairy cattle. Journal of Dairy Science. 2018;**2018**(101):872-888

[133] Chang G, Zhyang S, Seyfert H, Zhang K, Xu T, Jin D, et al. Chang, G.; Zhuang, S.; Seyfert, H.M.; Zhang, K.; hepatic TLR4 signaling is activated by LPS from digestive tract during SARA, and epigenetic mechanisms contribute to enforced TLR4 expression. Oncotarget. 2015;**6**:38578-38590

[134] Khafipour E, Krause DO, Plaizier JC. A grain-based subacute ruminal acidosis challenge causes translocation of lipopolysaccharide and triggers inflammation. Journal of Dairy Science A. 2009;**92**:1060-1070

[135] Xu Z, Regenstein JM, Xie D, Lu W, Ren X, Yuan J, et al. The oxidative stress and antioxidant rresponse of Litopenaeus vannamei to low temperature and air exposure. Fish Shellfish Immunology. 2018;**72**:564-571

[136] Zhao C, Liu G, Li X, Guan Y, Wang Y, Yuan X, et al. Inflammatory mechanism of Rumenitis in dairy cows with subacute ruminal acidosis. BMC Veterinary Research. 2018;**14**:1-8

[137] Montuschi P, Barnes PJ, Roberts LJ. Isoprostanes: Markers and mediators of oxidative stress. The FASEB Journal. 2004;**18**(15):1791-1800

[138] Bertoni G, Trevisi E, Han X, Bionaz M. Effects of inflammatory conditions on liver activity in puerperium period and consequences for performance in dairy cows. Journal of Dairy Science. 2008;**91**:3300-3310

[139] Yuan K, Farney JK, Mamdova LK, Sordllo LM, Bradford BJ. TNFalpha altered inflammatory responses, impaired health and productivity, but did not affect glucose or lipid

metabolism in earlylactation dairy cows. PLoS One. 2013;**8**:e80316

[140] Wang Y, Chun OK, Song WO. Plasma and dietary antioxidant status as cardiovascular disease risk factors: A review of human studies. Nutrients. 2013;**5**:2969-3004

[141] Urban-Chmiel R, Kankofer M, Werniki A, Albera E, Puchalski A. The influence of different doses of α-tocopherol andascorbic acid on selected oxidative stress parameters in in vitro culture of leukocytes isolated from transported calves. Livestock Science. 2009;**124**:89-92

[142] Chirase N, Greene L, Purdy R, Loan W, Auvermann B, Parker B, et al. Effect of transport stress on respiratory disease serum antioxidant status, and serum concentrations of lipid peroxidation biomarkers in beef cattle. American Journal of Veterinary Research. 2004;**65**:860-864

[143] Eitam H, Vaya J, Brosh A, Orlov A, Khatib S, Izhaki I, et al. Differential stress responses among newly received calves: Variations in reductant capacity and Hsp gene expression. Cell Stress & Chaperones. 2010;**15**:865-876

[144] Knowles TG, Warris PD, Vogel K. Stress physiology of animals during transport. En T. Gradin. Livestock Handling and Transport. 4th ed. Oxfordshire, UK: CABI; 2014. pp. 399–420

[145] Schwartzkopf-Genswein K, Ahola J, Edwards-Callaway D, Hale D, Paterson J. Transportation issues af fecting cattle well-being and considerations for the future. The Professional Animal Scientists. 2016;**32**:707-716

[146] Ismail IB, Al-Busadah KA, El-Bahr SM. Oxidative stress biomarkers and biochemical profile in broilers chicken fed zinc bacitracin and ascorbic acid under hot climate. American Journal of Biochemistry and Molecular Biology. 2013;**3**:202-214

[147] Panda A, Cherian G. Role of vitamin E in counteracting oxidative stress in poultry. The Journal of Poultry Science. 2013;**25**:1-28

[148] Ahmat H, Tian J, Wang J, Khan M, Wang Y, Zhang L, et al. Effects of dietary sodium selenite and selenium yeast on antioxidant enzyme activities and oxidative stability of chicken breast meat. Journal of Agricultural and Food Chemistry. 2012;**60**(29): 7111-7120

[149] Delles RM, Youling L, Xiong A, True D, Ao T, Dawson KA. Dietary antioxidant supplementation enhances lipid and protein oxidative stability of chicken broiler meat through promotion of antioxidant enzyme activity. Poultry Science. 2014;**93**(6):1561-1570

[150] Abdela N. Sub-acute ruminal acidosis (SARA) and its consequence in dairy cattle: A review of past and recent research at global prospective. Achievements in the Life Sciences. 2016; **10**(2):187-196

[151] Bernabucci U, Ronchi B, Lacetera N, Nardone A. Influence of body condition score on relationships between metabolic status and oxidative stress in Periparturient dairy cows. Journal of Dairy Science. 2005;**88**(6): 2017-2026

[152] Gheise JE, Riasi A, Zare A, Celi P, Ghoreishi SM. Effect of pre-calving body condition score and previous lactation on BCS change, bloodmetabolites, oxidative stress and milk production in Holstein dairy cows. Italian J of Animal Sci. 2017; **16**(3):474-483

[153] Colakoglu H, Yazlik M, Kurt S, Bayramoglu R, Vural M, Küplü Lü S, et al. Çolakoğlu, Hatice Yazlik, Murat relationship between body condition score, blood metabolites and oxidative stress in transition period and reproductive performance of dairy cows. Journal of Hellenic Veterinary Medical Society. 2020;**71**: 2041-2046

[154] Zhao W, Chen X, Xiao J, Chen X, Zhang X, Wang T, et al. Prepartum body condition score affects milk yield, lipid metabolism, and oxidation status of Holstein cows. Asian-Australasian Journal of Animal Sci. 2019;**32**(12): 1889-1896

[155] Bucktrout RM, Ma NA, Abdulrahman AA. One-carbon, carnitine, and glutathione metabolism-related biomarkers in peripartal Holstein cows by prepartal body condition. Journal of Dairy Science. 2020;**104**: 3403-3417. DOI: 10.3168/jds.2020-19402

[156] de Andrade KQ, Moura FA, dos Santos JM, de Aráujo OR, Santos JC, Goulart MO. Oxidative stress and inflammation in hepatic diseases: Therapeutic possibilities of N-acetylcysteine. International Journal of Molecular Sciences. 2015;**16**(12): 30269-30308

[157] Li S, Hong M, Tan HY, Wang N, Feng Y. Insights into the role and interdependence of oxidative stress and inflammation in liver diseases. Oxidative Medicine and Cellular Longevity. 2016; **2016**:4234061

[158] Mavrommatis A, Mitsiopoulou C, Christodoulou C, Kariamps P, Simoni M, Righi F, et al. Effects of supplementing rumen-protected methionine and lysine on Milk performance and oxidative status of dairy ewes. Antioxidants. 2021; **2021**(10):654

[159] Lopreiato V, Vailati-Riboni M, Belingeri A, Khan I, Farina G, Parys C, et al. Inflammation and oxidative stress transcription profiles due to in vitro supply of methionine with or without choline in unstimulated blood polymorphonuclear leukocytes from lactating Holstein cows. Journal of Dairy Science. 2019;**102**(11):10395-10410

Chapter 3

Acute Changes in Lipoprotein-Associated Oxidative Stress

Ngoc-Anh Le

Abstract

As inflammatory and oxidative stress are associated with cardiometabolic diseases, detection of abnormal fasting levels of inflammatory and oxidative biomarkers are indicative disease presence and may be too late for any preventive management. Metabolic flexibility refers to the ability of various metabolic processes to compensate for these acute changes and return all metabolites to baseline levels. By monitoring responses of key biomarkers to a standardized physiologic challenge, it is possible to assess the ability of the body to restore homeostasis, that is a measure of metabolic flexibility. Acute changes in lipoprotein-associated biomarkers of oxidative stress have been demonstrated following meal consumption. These include changes in circulating levels of oxidized low-density lipoproteins (LDL), levels of autoantibodies to malondialdehyde-modified LDL, as well as the oxidative susceptibility of isolated plasma LDL. These responses depend on the type and amount of dietary fats in the meal. Management with certain lipid-lowering drugs could also be shown to affect these meal-induced changes. However, plasma levels may be underestimated as we can demonstrate a spike in lipoprotein-associated biomarkers of oxidative stress resulting from the release oxidatively modified epitopes from the arterial wall by an intravenous bolus of heparin.

Keywords: lipoproteins, low-density lipoproteins, autoantibodies, MDA-LDL postprandial, metabolic flexibility, heparin

1. Introduction

Cardiometabolic diseases (CMD), specifically cardiovascular diseases and type 2 diabetes mellitus account for 17.5 and 1.5 million annual deaths, respectively [1, 2]. These noncommunicable diseases arise from metabolic abnormalities that should be preventable, especially if they can be identified at early stages. Current understanding would suggest that the onset and progression of CMD are the result of cumulative disturbances in cardiometabolic pathways [3]. Ongoing clinical diagnoses are typically based on cut-off values of various metabolites that have been associated with advanced diseases. Although these cut-off values have been established by major studies in a large number of participants, they are typically measured after

8 h of fasting. Since a typical individual may be consuming 2–3 meals daily, it is evident that metabolic processes are carried out in a non-fasting state throughout the day. It has been estimated that an individual adhering to the typical western diet may have triglyceride levels above fasting levels throughout 75% of the day [4]. Furthermore, some individuals may be exposed to a significant level of physical activity as part of their daily routine. Both meal consumption [5, 6] and physical activity [7, 8] are associated with transient spikes in oxidative stress. Indeed, while regular physical activity is believed to be associated with health benefits, excessive exercise has been reported to result in elevated biomarkers of oxidative damage in skeletal muscle and blood [8].

Metabolic flexibility or phenotypic flexibility is a term used to describe a battery of metabolic/physiologic processes that allow the body to regain homeostasis after various types of physiologic stresses [9, 10]. For instance, the oral glucose tolerance test is a widely accepted physiologic challenge to identify individuals with disturbed glucose metabolism (i.e. impaired flexibility) despite having normal levels of fasting glucose [11]. High-fat challenge tests have also been used to study lipid metabolism and metabolic flexibility [12]. The mode and intensity of acute exercise have also been reported to result in transient changes in several markers of chronic systemic inflammation, including C-reactive protein (hsCRP) and interleukin 6 (IL-6) [13].

While regulated oxidative stress is necessary for normal cellular metabolism, imbalance between reactive oxygen species (ROS) generation and ROS scavenging is recognized as a central step in the initiation and progression of CMD. However, direct measurement of ROS in the circulation is not feasible in view of their short biological half-lives which is of particular importance as all substrates in the body are potential targets of oxidative modification caused by ROS. A prime example of oxidation targets is polyunsaturated fatty acids (PSF) which are characterized by the multiple double bonds in their hydrocarbon chain [14]. Among other potential circulating biomarkers, plasma lipoproteins, and in particular, triglyceride-rich lipoproteins, come in direct contact with endothelial cells along the vasculature as part of their normal function to transport lipid throughout the body [15]. Free fatty acids, a key by-product of triglyceride metabolism, are highly susceptible to oxidative modification, move readily in and out of the sub-endothelium and are taken up by various tissues. Oxidative modification of lipids results in the formation of reactive aldehydes such as malondialdehyde (MDA) and 4-hydroxynonenal (HNE) [16]. MDA and HNE-modified metabolites can be detected in plasma [17].

2. Metabolism of triglyceride-rich lipoproteins and free fatty acids

Triglycerides (TG) and their structural component free fatty acids (FFA) represent the key source of energy in humans. In the fasting state, triglycerides are continuously synthesized and secreted by the liver after packaging in very-low-density lipoprotein (VLDL) particles for distribution and storage throughout the body. VLDL unload their cargo of triglycerides by attaching themselves to the endothelium with the subsequent release of free fatty acids and monoglycerides. The neutral and oxidized FFA released by the hydrolysis of triglyceride-rich VLDL can induce endothelial inflammation [18]. In the same process, any excess ROS generated by activated macrophages in the arterial wall could diffuse through the endothelium and seed lipoproteins while they are still in circulation [19]. The residual product after the delivery of TG is the cholesterol-rich low-density lipoprotein (LDL) particles (**Figure 1A**).

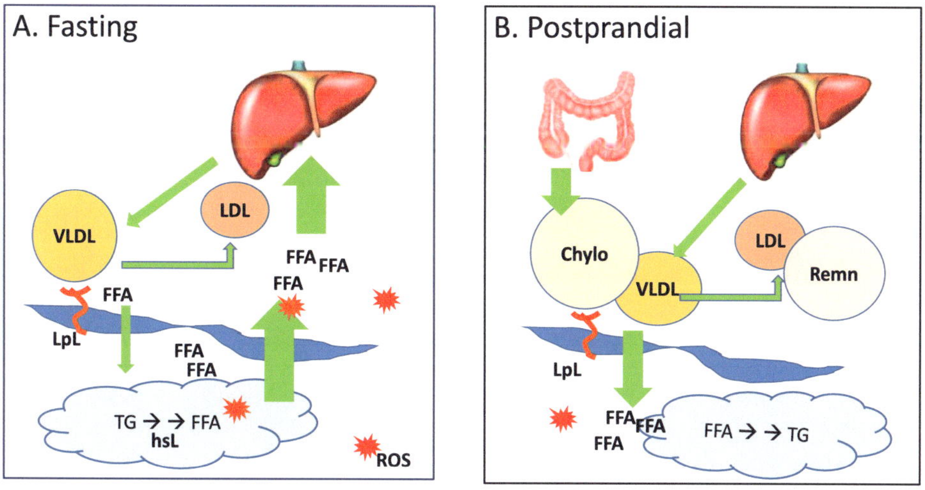

Figure 1.
Metabolism of triglyceride-rich lipoproteins (TRL) during fasting and postprandial state. TRL are attached to lipoprotein lipase via heparan sulfate proteoglycans allowing free fatty acids (FFA) to move in and out of the vascular space. Under inflammatory conditions, excess reactive oxygen species generated could diffuse to the lumen and potentially seed the surface of TRL while they are attached to the vessel wall. Without adequate anti-oxidant protection, the affected lipoproteins will undergo oxidative modification and thus acting as a biosensors of oxidative stress in the extravascular space.

In addition to the daily flux of endogenous TG secreted by the liver in the form of VLDL, approximately 90 g of fat are absorbed each day and packaged by the intestine within chylomicrons (Chylo) to transport newly ingested dietary fat [4]. For individuals with normal fasting TG levels (< 1.70 mmol/l, < 150 mg/dl), postprandial TG may be increased by 1.20 mmol/l (106 mg/dl), or approximately 70% [20]. VLDL and Chylo are commonly referred to as TG-rich lipoproteins (TRL).

Three key enzymes are responsible for the metabolism of circulating TG and FFA. They include heparin-releasable lipases, lipoprotein lipase (LpL) and hepatic lipase (HL), and hormone-sensitive lipase (hsL). Both TRL (VLDL and Chylo) share a common pathway requiring LpL-induced conversion of TG into non-esterified fatty acids (free fatty acids or FFA), that are taken up by cells and re-assembled as TG for storage in various tissues [15]. HL is primarily responsible for the breakdown of TG from the partially hydrolyzed TRL [15]. During fasting, when plasma insulin level is low, hsL is actively breaking down stored TG with the release of FFA back into the circulation to be transported back to the liver by albumin [21]. During postprandial lipemia, hsL is inhibited and there is a net increase in LpL-induced FFA release from Chylo for storage [21]. These metabolic processes are schematized in **Figure 1**.

LpL and HL are anchored to the luminal surface of vascular endothelial cells via membrane-bound heparan sulfate proteoglycans and could be released into the bloodstream by intravenous heparin [22]. Plasma TG levels are significantly reduced by as much as 80% following the release of the lipases by heparin [23, 24]. This acute reduction in plasma TG was explained by the failure of released FFA to inhibit LpL activity in the circulation [25]. LpL was originally described as the "clearing factor" [26, 27] for its ability to clear the turbidity due to elevated TG in plasma.

The residual product following the hydrolysis of VLDL by LpL is the cholesterol-rich LDL [28] that has been implicated in the initiation and progression of atherosclerosis [29]. According to the oxidation hypothesis of atherosclerosis [30, 31], only oxidatively modified LDL can be taken up by macrophages subsequently leading to the formation of foam cells, endothelium dysfunction, and the development and/or progression of atherosclerosis [32, 33]. As such, research on lipoprotein-associated oxidative stress has predominantly focused on oxidatively modified LDL. Starting with the detection of oxidatively modified LDL in atherosclerotic lesions [34, 35] and the evidence for an in vivo process for LDL oxidative modification [36], a number of studies linking circulating levels of oxidized LDL to cardiovascular disease emerged [37–39]. However, there is also increasing evidence that oxidative modification of lipoproteins is not limited to LDL but can affect other lipoproteins [40] with the susceptibility to oxidative modification being dependent on their fatty acid composition [41, 42]. We subsequently reported that the oxidative susceptibility of plasma LDL could be affected acutely following meal consumption [43].

3. Lipoprotein-associated oxidative stress

3.1 Autoantibodies against malondialdehyde-modified lipoproteins

While the presence of malondialdehyde-modified LDL (MDA-LDL) [38, 39, 44, 45] and autoantibodies against oxidatively modified LDL (AAb-oxLDL) [46–49] has been well accepted, the relationship between these biomarkers and atherosclerosis remains unclear. High levels of oxLDL have been reported to be associated with the presence and severity of CAD [17, 35, 38, 39, 44]. However, in spite of angiographically documented improvement in atherosclerosis with aggressive lipid-lowering intervention, plasma levels of oxLDL were actually increased [50]. Similarly, the association between AAb-oxLDL and atherosclerosis is not clear. In LDL-receptor deficient mice, levels of AAb-oxLDL positively correlated with atherosclerosis progression [51]. In the Watanabe rabbit lacking LDL receptor, these same investigators reported protection from atherosclerosis when levels AAb-oxLDL were increased by immunization with MDA-LDL [52]. Contradictory results have also been reported in humans with levels of AAb-oxLDL being associated with disease in some studies [46, 53, 54] but not in others [48, 49].

According to the oxidation hypothesis of atherosclerosis, only the smaller LDL particles could go through the endothelium, be trapped in the sub-endothelium, and undergo oxidative modification [30]. oxLDL would be in equilibrium between the endothelium and vascular space. An alternate scenario would allow plasma lipoproteins rich in highly oxidizable polyunsaturated fatty acids to attach to the arterial wall and be exposed to ROS generated in the sub-endothelium. Depending on the metabolic state preceding blood collection, levels of lipoprotein-associated biomarkers may be transiently affected resulting in different relationships between biomarkers and disease status. We have designed four different studies to examine this hypothesis.

In the first study, we administered a standardized mixed meal in the form of a 600 kcal shake to a group of patients with documented CAD and a small group of young healthy volunteers with no known risk factor for CAD [55]. Following the consumption of the shake, we observed a significant and transient reduction in plasma levels of autoantibodies (AAb) specific to MDA-LDL. This acute change was demonstrable only in individuals with documented CAD and not in healthy volunteers [55]. We have interpreted this reduction in AAb as suggestive of acute meal-induced increase in the levels of oxidatively modified lipoproteins. Thus, the time of the last meal prior to blood collection and the clearance of the intestinal lipoproteins would affect the levels of AAb-oxLDL.

In the second study, we examine the effect of different fatty acid compositions in the shake. In a small group of patients with metabolic syndrome, we examined the postprandial response with four different oral challenges administered in random order. The oral challenges were either enriched in different fatty acids, including polyunsaturated fatty acids (PSF), monounsaturated fatty acids (MSF), saturated fatty acids (SF), whereas one contained no fat (glucose). Details on the preparation of the shake have been presented elsewhere [56]. As expected we reported that this meal-induced reduction in AAb against MDA-LDL was specific for test meals rich in highly oxidizable polyunsaturated fatty acids (PSF) and could not be observed with challenges containing saturated fatty acids (SF), monounsaturated fatty acids (MSF), or only glucose (**Figure 2**) [56]. Thus, the composition of the last meal prior to blood collection could have an impact on plasma levels of AAb-oxLDL.

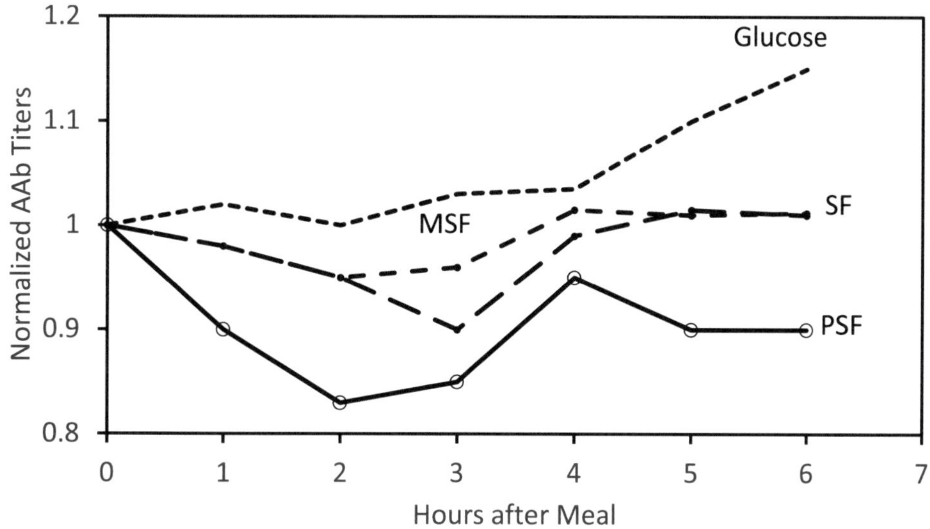

Figure 2.
Meal-induced changes in levels of AAb against MDA-LDL: Effect of meal composition. Following an oral challenge, there was a transient reduction in the circulating levels of autoantibodies (AAb) to malondialdehyde-modified lipoproteins. The acute change was observed only with a challenge enriched in the highly oxidizable polyunsaturated fatty acids (PSF) and not with monounsaturated fatty acids (MSF), saturated fatty acids (SF), or glucose.

In the third study, we examine the effect of therapy with a lipid-lowering drug known to have antioxidant properties on this postprandial response. For this we chose simvastatin, a well-known cholesterol-lowering drug that has previously been reported to have antioxidant properties [57], In a group of hypercholesterolemic patients we examined the postprandial response at baseline and after four months of treatment with simvastatin (40 mg/day). The composition of the test meal and laboratory analyses have been previously described [55]. We observed no difference in fasting levels of AAb against MDA-LDL (**Figure 3A**) despite a 30% reduction in LDL. However, the meal-induced reduction in AAb against MDA-LDL was completely blunted (**Figure 3B**) indicative of a reduction in oxidative stress in the arterial wall with simvastatin therapy. This result would suggest that the oxidative state of the arterial wall could have an impact on plasma levels of AAb-oxLDL.

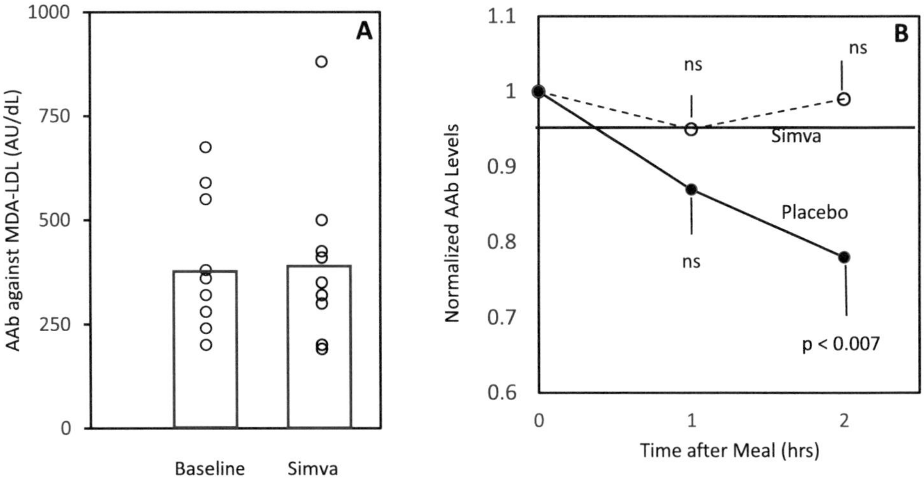

Figure 3.
Effect of simvastatin therapy on levels of autoantibodies against MDA-LDL A. Following 4-month treatment with simvastatin (40 mg/day), there was no difference in the levels of AAb to MDA-LDL in fasting plasma. B. At 2 h following the consumption of a mixed composition shake enriched in polyunsaturated fatty acids, the 22% reduction in AAb levels observed during the placebo period could not be demonstrated after simvastatin management.

In the fourth study, we used an intravenous bolus of heparin (60 IU/kg body weight) to dislodge metabolites that would be attached to heparan sulfate proteoglycans lining the arterial wall (unpublished data, pre- and post-heparin plasma samples were made available courtesy of Dr. RB Goldberg and colleagues at the University of Miami).

In addition to the measurements of AAb against MDA-LDL and LDL-IgG immune complexes as previously described [55] we also measured levels of oxLDL and myeloperoxidase in pre-and post-heparin samples by immunoassay (Mercodia, Winston-Salem, NC, USA). MPO is an enzyme released by leukocytes with the specific task of producing reactive oxidants for the destruction of ingested microorganisms within phagosomes [58, 59] and has been linked to atherosclerosis [60]. MPO is also recognized as a key catalyst for lipoprotein oxidation [61]. Concentrations of apoB and apoA-I, the key structural protein in TRL + LDL and HDL, respectively, were also measured by immunoturbidometric method (Sekisui Diagnostics, Burlington, MA, USA) on the AU480 automatic chemistry analyzer (Beckman Diagnostics, Brea, CA, USA).

Figure 4A illustrates acute reductions in plasma TG by 59.1% and 55.8% for individuals with normal glucose tolerance test (NGT) as compared to individuals with impaired oral glucose tolerance test (IGT + DM), respectively. Concomitant with this reduction in plasma TG, circulating levels of oxidatively modified LDL (oxLDL) were significantly increased, 27% and 23% ($p < 0.001$) in NGT and IGT + DM, respectively. Plasma levels of AAb against MDA-LDL (15.6% and 21.8%) and LDL-IgG immune complexes (21.9% and 16.9%) were also increased, independent of diabetes status ($p < 0.001$). We also note a significant increase (9.4% and 9.3%, $p < 0.001$) in plasma levels of myeloperoxidase (MPO) by immunoassay. There was no difference in levels of CHOL, apoB, or apoA-I between the pre- and post-heparin plasma samples (**Figure 4B**). Thus, lipoprotein-associated markers released into the circulation by the bolus of heparin were preferentially oxidatively modified. This is direct evidence that levels of oxidatively modified biomarkers in plasma may be underestimated due to

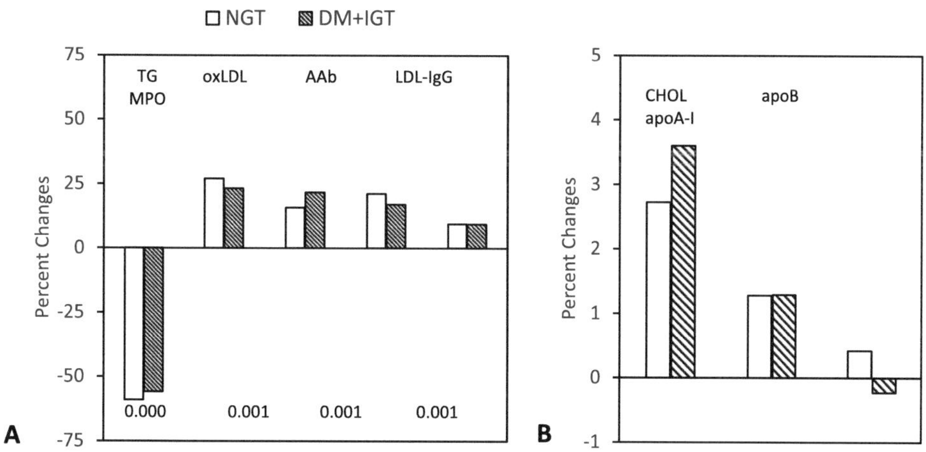

Figure 4.
Acute effects of intravenous heparin. A. At 15 min following a bolus intravenous injection of heparin to release lipoprotein lipase from its heparan sulfate proteoglycan anchor, there is a significant reduction in plasma TG concomitant with significant increase in plasma levels of oxLDL, AAb against MDA-LDL, IDL-IgG immune complexes and MPO. There was no difference in response whether the participants have a normal glucose tolerance test (NGT, open bars) or impaired test (DM + IGT, hatched bars). B. Intravenous heparin has no effect on plasma levels of cholesterol, apoB (key protein on TRL and LDL) or apoA-I (key protein on high density lipoproteins).

retention along the arterial wall of these metabolites. Depending on the propensity of oxidatively modified lipoproteins to adhere to the arterial wall, plasma levels of both oxLDL and AAb-oxLDL may not reflect the true levels of these biomarkers.

3.2 Lipoprotein oxidative susceptibility

Considering the abundance of both nonenzymatic (glutathione, uric acid, bilirubin, vitamin E, vitamin C, etc.) and enzymatic antioxidants (superoxide dismutase, glutathione reductase, catalase, etc.) in the body, the mechanism(s) by which an acute increase in oxidatively modified lipoproteins would occur is unclear. The concept of oxidative susceptibility may be helpful to explain this observation. Oxidative susceptibility of isolated lipoproteins is based on the kinetics of the formation of conjugated dienes when exposed to a highly oxidative environment. By exposing the preparation of purified lipoproteins to Cu^{++} as an artificial catalyst of oxidation, Esterbauer et al. [62] characterized the oxidative process in terms of changes in absorbance at 234 nm, corresponding to absorption by newly formed dienes. They defined the oxidative process in three phases, lag phase corresponding to the initial formation of dienes in the presence of endogenous antioxidants, propagation phase associated with the accelerated formation of dienes after antioxidant properties have been depleted, and decomposition phase. As endogenous antioxidants are consumed by initiating free radical species during the lag phase, lipoprotein particles that are more susceptible to oxidative modification would have shorter lag phase.

Increased oxidative susceptibility of plasma lipoproteins has been linked to CAD [63], stroke [64], diabetes mellitus [41] as well as kidney disease [65] among other chronic conditions. Oxidative susceptibility of LDL has also been shown to be affected by diet [66–68] and physical activity [69]. Data from our group indicate that lipoprotein susceptibility is acutely affected by meal consumption and can be altered by lipid-lowering therapy.

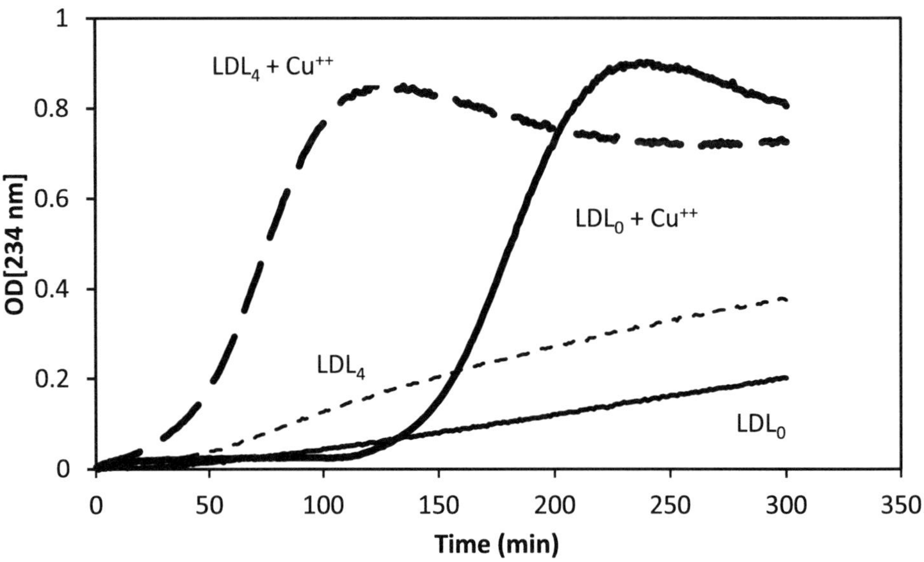

Figure 5.
Effect of postprandial lipemia on susceptibility to oxidative modification of fasting and postprandial LDL. In the absence of the oxidant catalyst Cu++, the rate of formation of conjugated dienes is minimal for both LDL isolated from fasting plasma (LDLo, light solid line) or from plasma collected at 4 hr (LDL4, short dash line) after meal consumption. In the presence of Cu++, there was generation of conjugated dienes with the rate being significantly faster with postprandial LDL (LDL4, long dash line) than with fasting LDL (LDLo, heavy solid line).

Figure 5 illustrates the time-dependent formation of conjugated dienes (as assessed by optical density at 234 nm) for preparations of LDL isolated at t = 0 h (LDL_0) and at t = 4 h (LDL_4) following meal consumption. In the absence of Cu^{++} as a catalyst, there was minimal formation of conjugated dienes with either the fasting LDL or the postprandial LDL samples. Upon incubation with Cu^{++}, there was rapid formation of conjugated dienes with the lag time of postprandial LDL (LDL_4) being significantly shorter than that of fasting LDL (LDL_0), 38.7 min versus 142.4 min [43], respectively. We hypothesize that, in individuals with metabolic syndrome, postprandial lipoproteins might be seeded with ROS during their contact with the arterial wall and therefore could be more susceptible to oxidative modification when compared to fasting lipoproteins.

Figure 6 presents changes in oxidative susceptibility of fasting and postprandial LDL following a 6-month treatment with a TG-lowering medication (ABT-335, Trilipix[c]) in a patient with metabolic syndrome. As shown (**Figure 5A**), in the absence of Cu^{++}, fasting LDL from this participant underwent spontaneous oxidative modification. The rate of formation of conjugated dienes was significantly slowed down after ABT-335 therapy. In the presence of Cu^{++}, the lag time of fasting LDL was increased from 28.7 min to 149.5 min after therapy. The lag time for postprandial LDL was minimally increased after therapy, 25.5 min versus 35.5 min, respectively [43].

3.3 oxLDL versus LDL

A more direct assessment of the meal-induced in oxidation biomarkers is the direct measurement of oxLDL and total plasma apoB by immunoassay following meal consumption (unpublished data). Apolipoprotein B (apoB) is the primary structural

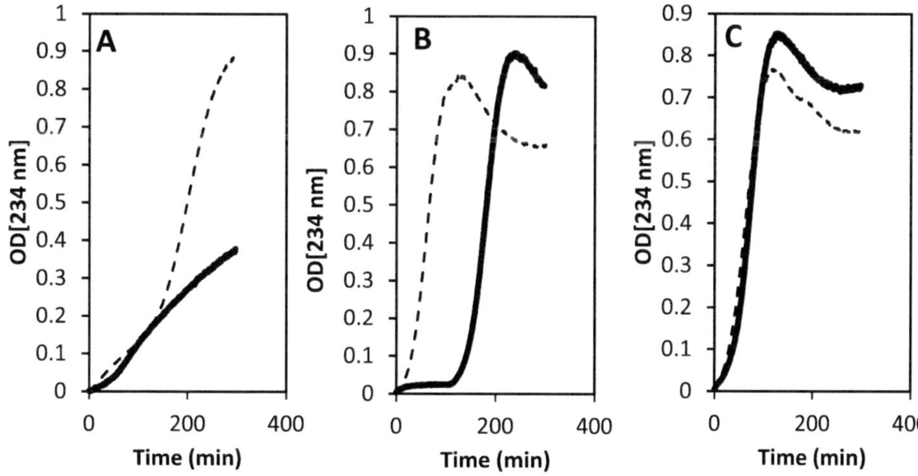

Figure 6.
Effect of ABT-335 on susceptibility to oxidative modification of fasting and postprandial LDL. A. For an individual with metabolic syndrome, fasting LDL undergo spontaneous oxidative modification as demonstrated by the generation of conjugated dienes in the absence of Cu^{++} (dashed line). After a 6-month therapy with ABT-335, the rate of spontaneous oxidative modification of fasting LDL is significantly reduced (solid line). B. In the presence of Cu^{++}, the rate of formation of conjugated dienes for fasting LDL was approximately 4-fold during the placebo period (dashed line) as compared to after treatment with ABT-335 (solid line). C. In the presence of Cu^{++}, the rate of formation of conjugated dienes for postprandial LDL was not affected by the treatment with ABT-335.

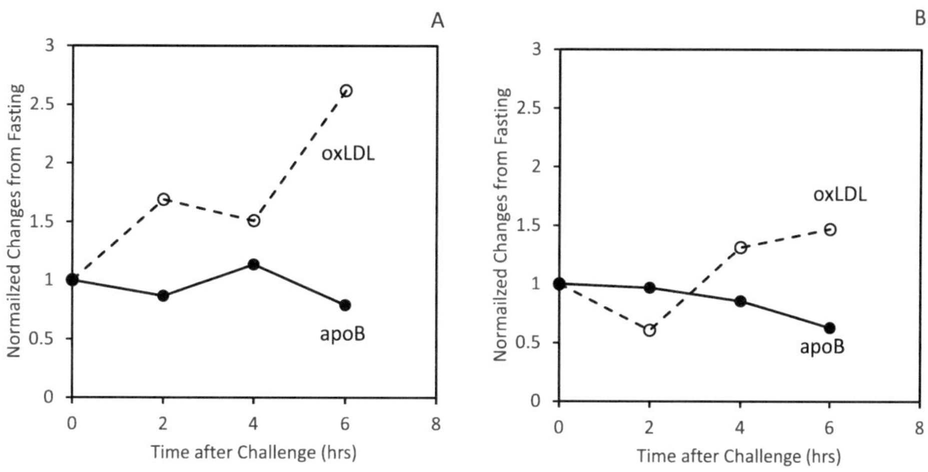

Figure 7.
Postprandial changes in levels of oxLDL and apoB A. Following the oral challenge, there is minimal change in total plasma apoB while the plasma level of oxLDL increase by 50% after 4 hrs and by 250% by 6 h. B. In this individual, the meal-induced change in oxLDL level is more modest with only a 150% increase by 6 h. Total plasma apoB remains minimally affected by the oral challenge.

protein in TRL and LDL. In a preliminary study, we examine the change in plasma levels of oxLDL (Mercodia, Winston-Salem, NC, USA) and apoB (Sekisui Diagnostics, Burlington, MA, USA). **Figure 7** illustrates representative time-dependent changes in these metabolites following the standardized oral challenge for two participants. In both individuals, there was minimal change in total plasma apoB following the meal.

Plasma oxLDL levels were increased in both instances, reaching a maximal increase of 2.5-fold (**Figure 7A**) and 1.5-fold (**Figure 7B**) 6 h after meal consumption. While apoB and oxLDL levels have been examined previously this is the first data suggesting that they are not affected to the same degree during postprandial lipemia. Further investigation is needed to confirm this observation and to determine when oxLDL levels return back to pre-meal levels.

4. Metabolic flexibility

There has been a recent shift away from the traditional definition of health formulated by the WHO in 1948 as "a state of complete physical, mental, and social well-being and not merely the absence of disease or infirmity" [70]. The current concept of health is based on the definition by Huber et al. [70] in terms of flexibility, adaptability, elasticity, and robustness. In other words, health should be defined in terms of adaptation and flexibility to continuously changing external inputs [71]. This ability to adapt to a specific metabolic challenge is referred to as metabolic flexibility or phenotypic flexibility [10, 71].

The most commonly used challenge test in the clinical environment is the oral glucose tolerance test which is the primary tool for the diagnosis of diabetes [72]. Ceriello et al. [5, 6, 73–75] have published extensively on the acute effect of oral glucose administered by itself or as part of a mixed meal on biomarkers of inflammation and oxidative stress. Van Ommen and his colleagues have characterized a number of metabolic processes that could be associated with clusters of metabolites using their "PhenFlex challenge" [9, 10, 71, 76–78].

The studies by both the Ceriello team and the Van Ommen team, though very comprehensive, utilized non-physiologic dosage of fat in their challenges, 75 grams [5] and 60 grams [76], respectively. Especially, since the typical consumption of a 600-kcal meal with 30% fat composition as recommended by the AHA would correspond to only 20 grams of fat. Furthermore, the PhenFlex challenge consisted primarily of monounsaturated fatty acids (47%) and saturated fatty acids (39%). The high content of monounsaturated fatty acids may blunt any inflammatory effects of dietary fat [79, 80]. The low contents of highly oxidizable polyunsaturated fatty acids (only 14% of the fat intake) on the other hand, might not be sufficient to elicit any lipoprotein-associated oxidative stress.

To date, most studies utilizing metabolic challenges have limited themselves to looking at the transient changes in the levels of various metabolites as a function of time. Since the metabolism of many of these metabolites is inter-related, we propose to examine phenotypic flexibility in terms of trajectory analysis using two metabolites that are under coordinated control. With an oral meal challenge, the simplest pair of metabolites is glucose and insulin. We used postprandial glucose and insulin data from a group of patients with type 2 diabetes mellitus participating in a randomized double-blind, multicenter clinical trial comparing the efficacy of mid-mixture lispro versus glargine [81]. **Figure 8** presents the postprandial changes in the mean levels of insulin and glucose as a function of time during glargine (Panel A) or MM-lispro (Panel B) therapy. In panel C (**Figure 8C**), each data point represents the normalized levels of insulin and glucose at a specific time following meal consumption. Both trajectories start out at (1,1) and return back to the starting point after 8 h. Under glargine (dashed), the maximum excursion for

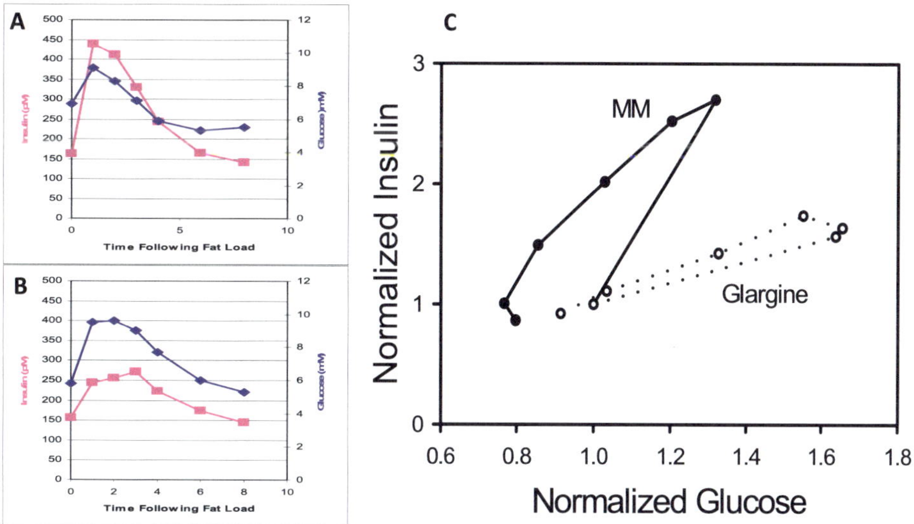

Figure 8.
Trajectory Analysis. Time-dependent changes in Insulin (filled squares) and glucose (filled diamonds) for a group of diabetic patients stabilized on glargine (Panel A) or MM (Panel B). Panel C: each data point represents the normalized value of glucose versus the normalized value of insulin at different times.

glucose at t = 2 h is at 1.7 (or 70% greater than fasting level) and at 1.5 for insulin. In contrast, with the administration of pre-meal lispro (solid line), the maximum excursion of glucose is 1.3 at t = 1 h corresponding to an insulin level of 2.7 which reflects the combined concentrations of endogenous and exogenous insulin. This trajectory analysis can be characterized by several parameters, including the maximum excursion, slope of the maximum excursion, perimeter of the trajectory, and area enclosed within the trajectory. A large maximum excursion distance would suggest great impact of the challenge while a large perimeter and enclosed area would be indicative of poor metabolic control.

5. Conclusions

CMD are chronic conditions associated with inflammatory and oxidative stress and tend to progress over time. Abnormal fasting levels of inflammatory and oxidative biomarkers are typically indicative of presence of advanced disease conditions and may be too late for any preventive management. The ability to assess how rapidly the body can respond to a challenge and maintain metabolic homeostasis may be useful in establishing early signs of metabolic disturbances. It is crucial to note that the challenge should be similar to physiological influences and must be reflective of typical daily activities. While the use of an oral meal challenge may be used to demonstrate the beneficial effects of some nutrients, the composition of the challenge must include elements known to elicit the maximal responses in terms of inflammatory and oxidative stress. Furthermore, in addition to examining the time-dependent changes, it would be helpful to identify biomarkers that are metabolically linked as that may provide data that could elucidate potential regulatory pathways.

Acknowledgements

We are indebted to the medical staff of the Lipid Research Program at the Atlanta Veterans Affairs Healthcare System for their support in recruiting and management of the participants and of the Biomarker Core Laboratory, in particular Dr. Monica Farkas-Epperson. The collaboration of Drs. R. Goldberg and H. Florez, the University of Miami for the post-heparin samples are greatly appreciated.

Author details

Ngoc-Anh Le
Biomarker Core Laboratory, Foundation for Atlanta Veterans Education and Research (FAVER), Decatur, Georgia, USA

*Address all correspondence to: anh.le@va.gov

References

[1] Benzinger CP, Roth GA, Moran AE. The global burden of disease study and the preventale burden of NCD. Global Heart. 2016;**11**(4):393-397

[2] Batakumar P, Maung U, Jagadeesh G. Prevalence and prevention of cardiovascular disease and diabetes mellitus. Pharmacological Reviews. 2016;**113**(Pt A):600-609

[3] Ndisang J, Rastogi S. Cardiometabolic diseasesand related complications: Current status and future perspective. Biomedical Research. 2013. DOI: 10.1155/2013/467682

[4] Keirns BH, Sciarrillo CM, Koemel NA, Emerson SR. Fasting, non-fasting and postprandial triglycerides for screening cardiometabolic risk. Journal of Nutritional Science. 2021;**10**:75-89

[5] Ceriello A, Taboga C, Tonutti L, Quagliaro L, Piconi L, Bais B, et al. Evidence for an independent and cumulative effect of postprandial hypertriglyceridemia and hyperglycemia on endothelial dysfunction and oxidative stress generation. Effects of short- and long-term simvastatin treatment. Circulation. 2002;**106**:1211-1218

[6] Ceriello A, Bortolotti N, Motz E, Crescentini A, Lizzio S, Russo A, et al. Meal-generated oxidative stress in type 2 diabetes. Diabetes Care. 1998;**21**:1529-1533

[7] Powers SK, Radak Z, Ji LL. Exercise-induced oxidative stress: Past, present and future. The Journal of Physiology. 2016;**18**:5081-5092

[8] Powers S, Jackson M. Exercise-induced oxidative stress: Cellular mechanisms and impact on muscle force production. Physiological Reviews. 2008;**88**:1243-1276

[9] van Ommen B, Keijer J, Kleemann R, Elliott R, Drevon CA, McArdle H, et al. The challenges for molecular nutrition research 2: Quatification of the nutritional phenotype. Genes & Nutrition. 2008;**3**(2):51-59

[10] van Ommen B, van der Greef J, Ordovas JM, Daniel H. Phenotypic flexibility as key factor in the human nutrition and health relationship. Genes & Nutrition. 2014;**9**:423

[11] Shaham O, Wei R, Wang T, Ricciardi C, Lewis G, Vasan R, et al. Metabolic profiling of the human response to a glucose challenge reveals distinct axes of insulin sensitivity. Molecular Systems Biology. 2008;**4**:214-223

[12] van Dijk S, Mensink M, Esser D, Feskens E, Muller M, Afman I. Responses to high-fat challenges varying in fat type in subjects with different metabolic risk phenotypes: A randomized trial. PLoS One. 2012;**7**:e41388

[13] Mendham A, Donges C, Liberts E, Duffield R. Effects of mode and intesnity on the acute exercise-induced IL-6 and CRP responses in a sedentary, overweight population. European Journal of Applied Physiology. 2010;**111**:1035-1045

[14] Niki E, Yoshida Y, Saito Y, Noguchi N. Lipid peroxidation: Mechanisms, inhibition, and biological effects. Biochemical and Biophysical Research Communications. 2005;**338**:668-676

[15] Goldberg I. Fat in the blood, fat in the artery, fat in the heart: Triglyceride in

physiology and disease. Arteriosclerosis, Thrombosis, and Vascular Biology. 2018;**38**:700-706

[16] Esterbauer H, Schaur RJ, Zollner H. Chemistry and biochemistry of 4-hydroxynonenal, malondialdehyde and related aldehydes. Free Radical Biology & Medicine. 1991;**11**:81-128

[17] Holvoet P, Perez G, Zhao Z. MDA-modified LDL in patients with atherosclerotic disease. The Journal of Clinical Investigation. 1995;**95**:2611-2619

[18] Wang L, Gill R, Pedersen TL, Higgins LJ, Newman JW, Rutledge JC. TG-rich lipoprotein lipolysis releases neutral and oxidized FFAs that induce endothelial cell inflammation. Journal of Lipid Research. 2009;**50**:204-213

[19] Le N-A. In: Waisundara VY, Jovandaric MZ, editors. Apolipoproteins, Triglycerides, and Cholesterol. IntechOpen: London UK; 2020. pp. 105-116

[20] Taguchi H, Watanabe H, Onizawa K, Nagao T, Gotoh N, Yasukawa T, et al. Double-blind controlled study on the effects of dietary diacylglycerol on postprandial serum and chylomicron triacylglycerol responses in healthy humans. Journal of the American College of Nutrition. 2000;**19**(6):789-796

[21] Alves-Bezerra M, Cohen D. Triglyceride metabolism in the liver. Comprehensive Physiology. 2019;**8**(1):1-8

[22] Baginsky ML, Brown WV. Differential chracteristics of purified hepatic triglyceride lipase and lipoprotein lipase from human postheparin plasma. Journal of Lipid Research. 1977;**18**(4):423-437

[23] Yang J, Kim T, Koo B, Park B, Park J. Change of plasma lipoproteins by heparin-released liporprotein lipase. Experimental & Molecular Medicine. 1999;**31**(2):60-64

[24] Jackson RL, Yates MT, McNerney CA, Kashyap ML. Relationship between post-heparin plasma lipases, triglycerides and high density lipoproteins in normal subjects. Hormone and Metabolic Research. 1990;**22**(5):289-294

[25] Brown WV, Baginsky ML. Some functional aspects of apolipoproteins: apoLp-Ala inhibition of lipoprotein lipase and deinhibition by monoolein. Hormone and Metabolic Research. 1974;**4**:11-16

[26] Spitzer JJ. Properties of heparin-prduced lipemia clearing factor. The American Journal of Physiology. 1952;**177**(2):492-298

[27] Korn ED. Clearing factor, a heparin-activated lipoprotein lipase: I. Ioslation and characterization of the enzyme from normal rat heart. The Journal of Biological Chemistry. 1955;**215**:1-14

[28] Sigurdsson G, Nicoll A, Lewis B. Conversion of very low density lipoprotein to low density lipoprotein. The Journal of Clinical Investigation. 1975;**56**:1481-1490

[29] Castelli W. Lipids, risk factors and ischemic heart disease. Atherosclerosis. 1996;**124**(Suppl):S1-S9

[30] Steinberg D, Parthasarathy S, Carew TE, Khoo JD, Witztum JL. Beyond cholesterol: Modifications of LDL that increase its atherogenecity. The New England Journal of Medicine. 1989;**320**:915-924

[31] Navab M, Anantharamaiah G, Reddy ST, van Lenten BJ, Ansell BJ, Fonarow GC, et al. The oxidation hypothesis of atherogenesis:

The role of oxidized phospholipids and HDL. Journal of Lipid Research. 2004;**45**:993-1007

[32] Stocker R. Oxidation of LDL with hypochlorite cause transformation of the lipoprotein into a high-uptake form for macrophages. The Biochemical Journal. 1993;**290**:165-172

[33] Hazell LJ, van den Berg JJ, Stocker R. Oxidation of LDL by hypochlorite cause aggregation that is mediated by modification of lysine residues rather than lipid oxidation. The Biochemical Journal. 1994;**302**:297-304

[34] Avogaro P, Bittolo Bon G, Cazzolato G. Presence of a modified LDL in humans. Arteriosclerosis. 1988;**8**:79-87

[35] Yla-Herttuala S, Palinski W, Rosenfeld ME, Parthasarathy S, Carew TE, Butler S, et al. Evidence for the presence of oxidatively modified low density lipoprotein in atherosclerotic lesions of rabbit and man. The Journal of Clinical Investigation. 1989;**84**(4):1086-1095

[36] Palinski W, Yla-Herttuala S, Gurtner GC, Socher SS, Butler SW, Parthasarathy S, et al. LDL undergoes oxidative modification in vivo. Proceedings of the National Academy Science USA. 1989;**86**:1372-1376

[37] Holvoet P, Kritchevsky SB, Tracy RP, Mertens A, Rubin SM, Butler J, et al. The metabolic syndrome, circulating oxidized LDL, and risk of myocardial infarction in well-functioning elderly people in the health, aging, and body composition cohort. Diabetes. 2004;**53**(4):1068-1073

[38] Holvoet P, Mertens A, Verhamme P, Bogaerts K, Beyens G, Verhaeghe R, et al. Circulating oxLDL is a useful marker for identifying patients with coronary artery disease. Arteriosclerosis, Thrombosis, and Vascular Biology. 2001;**21**:844-848

[39] Holvoet P, Jenny NS, Schreiner PJ, Tracy RP, Jacobs DR. The relationship between oxLDL and other cardiovascualr risk factors and subclinical CVD in different ethnic groups: The multi-ethnic study of atherosclerosis (MESA). Atherosclerosis. 2007;**194**:245-252

[40] Nakajima K, Nakano T, Tanaka A. The oxidative modification hypothesis of atherosclerosis: The comparison of atherogenic effects of oxidaized LDL and remnant lipoproteins in plasma. Clinica Chimica Acta. 2006;**367**:36-47

[41] McEneny J, O'Kane MJ, Moles KW, Mcmaster C, Mcmaster D, Mercer C, et al. VLDL subfractions in type II diabetes mellitus: Alterations in composition and susceptibility to oxidation. Diabetologia. 2000;**43**:485-493

[42] Mabile L, Salvayre R, Bonnafe MJ, Negre-Salvayre A. Oxidizability and subsequent cytotoxicity of chylomicrons to monocytic U937 and endothelial cells are dependent on dietary fatty acid composition. Free Radical Biology & Medicine. 1995;**19**:599-607

[43] Le N-A, Farkas-Epperson M, Sweeney ME, Wilson PWF, Brown WV. Effect of ABT-335 (fenofibric acid) on meal-induced oxidative stress in patients with metabolic syndrome. Atherosclerosis. 2013;**231**:268-273

[44] Wallenfeldt K, Fagerberg B, Wikstrand J, Hulthe J. OxLDL in plasma is a prognostic marker of subclinical atherosclerosis development in clinically healthy men. Journal of Internal Medicine. 2004;**256**:413-420

[45] Meisinger C, Baumert J, Khuseyinova N, Loewel H, Koenig W.

Plasma oxLDL, a strong predictor for acute coronary heart disease events in apparently healthy, middle-aged men from the general population. Circulation. 2005;**112**:651-657

[46] Bui MN, Sack MN, Moutsatsos G, et al. Autoantibody titers to OxLDL in patients with coronary atherosclerosis. American Heart Journal. 1996;**131**:663-667

[47] Maggi E, Finardi G, Poli M, Bollati P, Filipponi M, Stefano PL, et al. Specificity of autoantibodies against oxLDL predicting myocardial infarction. Coronary Artery Disease. 1993;**4**:1119-1122

[48] van de Vijver LPL, Steyger R, van Poppel G, et al. Autoantibodies against MDA-LDL in subjects with severe and minor atherosclerosis and healthy population controls. Atherosclerosis. 1996;**122**:245-253

[49] Uusitupa MI, Niskanen L, Luoma J, Vilja P, Mercuri M, Rauramaa R, et al. Autoantibodies against oxidized LDL do not predict atherosclerotic vascular disease in non-insulin-dependent diabetes mellitus. Arteriosclerosis, Thrombosis, and Vascular Biology. 1996;**16**(10):1236-1242

[50] Choi S, Chae A, Miller E, Messig M, Ntanios F, DeMaria A, et al. Relationship between biomarkers of oxLDL, statin therapy, quantitative coronary angiography, and atheroma volume. Journal of the American College of Cardiology. 2008;**52**(1):24-32

[51] Palinski W, Tangirala RK, Miller E, Young SG, Witztum JL. Increased autoantibody titers against epitopes of oxidized LDL in LDL-receptor deficient mice with increased atherosclerosis. Arteriosclerosis, Thrombosis, and Vascular Biology. 1995;**15**:1569-1576

[52] Palinski W, Miller E, Witztum JL. Immunization of LDL receptor-deficient rabbits with homologous malondialdehyde-modified LDL reduces atherosclerosis. Proceedings of the National Academy Science USA. 1995;**92**:821-825

[53] Puurunen M, Manttari M, Manninen V, et al. Antibody against oxidized LDL predicting myocardial infarction. Archives of Internal Medicine. 1994;**154**:2605-2609

[54] Mustafa A, Nityanand S, Berglund L, Lithell H, Lefvert AK. Circulating immune complexes in 50-year-old men as a strong and independent risk factor for myocardial infarction. Circulation. 2000;**102**(21):2576-2581

[55] Le N-A, Li X, Kyung S, Brown WV. Evidence for the in vivo generation of oxidatively modified epitopes in patients with documented CAD. Metabolism. 2000;**49**:1271-1277

[56] Gradek Q, Harris M, Yahia N, Davis W, Le N-A, Brown W. Polyunsaturated fatty acids acutely suppress antibodies to malondialdehyde-modified LDL in patients with vascular disease. The American Journal of Cardiology. 2004;**93**:881-885

[57] Girona J, La Ville AE, Sola R, Plana N, Masana L. Simvastatin decreases aldehyde production derived from lipoprotein oxidation. The American Journal of Cardiology. 1999;**83**(6):846-851

[58] Kettle AJ, WInterbourn CC. Myeloperoxidase: A key regulator of neutrophil oxidant production. Redox Report. 1997;**3**(1):3-15

[59] Klebanoff SJ. Myeloperoxidase: Friend and foe. Journal of Leukocyte Biology. 2005;**77**(5):598-625

[60] Koeth RA, Haselden V, Tang WH. Myeloperoxidase in cardiovascular disease. Advances in Clinical Chemistry. 2013;**62**:1-32

[61] Daugherty A, Dunn JL, Rateri DL, Heinecke JW. Myeloperoxidase, a catalyst for lipoprotein oxidation, is expressed in human atherosclerotic lesions. The Journal of Clinical Investigation. 1994;**94**(1):437-444

[62] Esterbauer H, Striegl G, Puhl H, Rotheneder M. Continuous monitoring of in vitro oxidation of human LDL. Free Radical Biology & Medicine. 1989;**6**:67-75

[63] Hendrickson A, McKinstry LA, Lewis JK, Lum J, Louie A, Schellenberg GD, et al. Ex vivo measures of LDL oxidative susceptibility predict carotid artery disease. Atherosclerosis. 2005;**179**:147-153

[64] Ryglewicz D, Rodo M, Roszczynko M, Baranska-Gieruszczak M, Szirkowiec W, Swiderska M, et al. Dynamics of LDL oxidation in ischemic stroke patients. Acta Neurologica Scandinavica. 2002;**105**:185-188

[65] McEneny J, Loughrey CM, McNamee PT, Trimble ER, Young IS. Susceptibility of VLDL to oxidation in patients on regular haemodialysis. Atherosclerosis. 1997;**129**:215-220

[66] Reaven P, Parthasarathy S, Grasse BJ, Miller E, Almazan F, Mattson FH, et al. Feasibility of using an oleate-rich diet to reduce the susceptibility of low-density lipoprotein to oxidative modification in humans. The American Journal of Clinical Nutrition. 1991;**54**(4):701-706

[67] Reaven P, Parthasarathy S, Grasse BJ, Miller E, Steinberg D, Witztum JL. Effects of oleate-rich and linoleate-rich diets on the susceptibility of LDL to oxidative modification in mildly hypercholesterolemic subjects. The Journal of Clinical Investigation. 1993;**91**:668-676

[68] Nielsen NS, Pedersen A, Sandstrom B, Marckmann P, Hoy CE. Different effects of diets rich in olive oil, rapeseed oil and sunflower-seed oil on postprandial lipid and lipoproteins concentrations and on lipoprotein oxidation susceptibility. The British Journal of Nutrition. 2002;**87**:489-499

[69] Parks EJ, German JB, Davis PA, Frankel EN, Kappagoda CT, Rutledge JC, et al. Reduced oxidative susceptibility of LDL from patients participating in an intensive atherosclerosis treatment program. The American Journal of Clinical Nutrition. 1998;**68**(4):778-785

[70] Huber M, Knottnerus JA, Green L, van der Horst H, Jadad AR, Kromhout D, et al. How should we define health? British Medical Journal. 2011;**343**:d4163

[71] Stroeve JHM, van Wietmarschen H, Kremer BHA, van Ommen B, Woperels S. Phenotypic flexibility as a measure of health: The optimal nutritional stress response test. Genes & Nutrition. 2015;**10**:13-34

[72] Group NDD. Classification and diagnosis of diabetes mellitus and other categories of glucose intolerance. Diabetes. 1979;**28**:1039-1057

[73] Esposito K, Nappo F, Marfella P, Giugliano G, Giugliano F, Ciotola M, et al. Inflammatory cytokine concentrations are acutely increased by hyperglycemia in humans: Role of oxidative stress. Circulation. 2002;**106**:2067-2072

[74] Ceriello A, Quagliaro L, Piconi L, Assaloni R, Da Ros R, Maier A, et al. Effect of postprandial hypertriglyceridemia and hyperglycemia on circulating adhesion molecules and oxidative stress generation and the possible role of simvastatin treatment. Diabetes. 2004;**53**:701-710

[75] Ceriello A, Quagliaro L, Catone B, Pascon R, Piazzola M, Bais B, et al. Role of hyperglycemia in nitrotyrosine postprandial generation. Diabetes Care. 2002;**25**:1439-1443

[76] Wopereis S, Stroeve JHM, Stafleu A, Bakker GCM, Burggraaf J, van Erk MJ, et al. Multi-parameter comparison of a standardized mixedmeal tolerance test in healthy and type 2 diabetic subjects: The PhenFlex challenge. Genes & Nutrition. 2017;**12**:21-35

[77] Wopereis S, Wolvers D, van Erk MJ, Gribnau M, Kremer B, van Dorsten FA, et al. Assessment of inflammatory resilience in healthy subjects using dietary lipid and glucose challenges. BMC Medical Genomics. 2013;**6**:44-60

[78] van den Broek TJ, Bakker GCM, Rubingh CM, Bijlsma S, Stroeve JHM, van Ommen B, et al. Ranges of phenotypic flexibility in healthy subjects. Genes & Nutrition. 2017;**12**:32-46

[79] Shroder H. Protective mechanisms of the Mediterranean diet in obesity and type 2 diabetes. The Journal of Nutritional Biochemistry. 2007;**18**(3):149-160

[80] Giugliano D, Esposito K. Mediterranean diet and metabolic diseases. Current Opinion in Lipidology. 2008;**19**(1):63-68

[81] Beisswenger PJ, Brown WV, Ceriello A, Le N-A, Goldberg RB, Cooke JP, et al. Meal-induced increases in C-reactive protein, IL6, and TNFa are attenuated by prandial + basal insulin in patients with type 2 diabetes. Diabetic Medicine. 2011;**28**:1088-1095

Chapter 4

The Importance of the Redox Modulation in the Prevention and Treatment of Chronic Pulmonary Diseases

Emma Borrelli

Abstract

This chapter discusses the most important mechanisms of action of oxidants in the pathogenesis of chronic pulmonary oxidative diseases and the possible use of redox modulators in the prevention and treatment of oxidant/antioxidant intracellular imbalance. Recent acquisitions on cellular physiology reported the key role, in micro-molecular doses, of reactive oxygen species (ROS) as signaling molecules although excessive ROS contribute to the development and progression of a large spectrum of diseases, including chronic obstructive pulmonary disease (COPD) and idiopathic pulmonary fibrosis (IPF). Therefore, a correct understanding of the roles of redox regulation in the respiratory system during the impairment of oxidative balance and the subsequent development of chronic lung diseases appears to be important. Moreover, an interdependence between oxidant and inflammatory mediators has been shown in several experimental studies on chronic lung diseases, making more intriguing the comprehension of the pathophysiological phenomena and the therapeutic approach. This chapter discusses the role of various exogenous substances targeting oxidant/antioxidant balance in the treatment of COPD and IPF and their very limited beneficial effects due to the reduced bioavailability in the human body. Finally, the importance of novel routes of administration or a combination of redox modulators will be discussed as a promising avenue for the prevention and treatment of this common and highly disabling disease.

Keywords: redox modulator, chronic lung disease, reactive oxygen species, oxidative stress, antioxidant

1. Introduction

Clinical and experimental studies suggest that oxidative stress (OS) and cellular redox balance have been implicated in the pathobiology of two major chronic lung diseases: chronic obstructive pulmonary disease (COPD) and idiopathic pulmonary fibrosis (IPF) [1]. Chronic obstructive pulmonary disease (COPD) is a pathological condition of the lung where both oxidative imbalance and inflammation play an important role in the development and progression of the disease [2]. In support of this hypothesis, a large series of scientific studies reported in COPD an elevated oxidant generation from environmental exposures, mainly tobacco smoke and air pollutants, associated with increased amounts of reactive oxygen species (ROS) and reactive nitrogen species (RNS) released from leukocytes and macrophages during the inflammatory process in the lungs [3]. The increased levels of oxidants may have multiple consequences, such as induction of proteases, inactivation of defense mechanisms, and activation of growth factors and inflammatory pathways [4]. More interestingly, COPD is widely recognized as not simply an inflammatory/destructive lung disease but also a chronic degenerative systemic disease with extrapulmonary manifestations, such as cardiovascular disease, skeletal muscle dysfunction, osteoporosis, and neurological degeneration [5, 6]. Thus, it appears clear that, in the lung cells of COPD patients, a vicious cycle of persistent inflammation accompanied by chronic oxidative stress begins locally but rapidly becomes a chronic challenge for the organism, if the process is not properly controlled or neutralized [7]. It therefore seems obvious that, to prevent the onset and the progression of this disease, the reduction of exogenous exposition to oxidants (e.g., smoking cessation), downregulation of endogenous ROS generation and inflammation, and an increase in antioxidants could represent the best therapeutic options [8–10]. As far as IPF, it has been ascertained that many exogenous substances, such as asbestos, radiation and drugs, are involved in the pathogenesis of pulmonary fibrosis through an augmented generation of ROS and a subsequent imbalance between oxidant and antioxidant substances in the lower respiratory tract [11, 12]. In bronchoalveolar lavage fluid of patients with IPF it has been found that an increased level of 8-isoprostane (a marker of oxidative stress) respects healthy patients [13]. Mitochondrial generation of ROS in epithelial lung cells is also increased in patients with IPF, and this phenomenon could lead to an increased apoptosis of the cells. The antioxidant system is reduced as suggested by a decrease in glutathione content in the lining fluid of the epithelial cells in IPF patients [14]. Unlike COPD patients, patients affected by idiopathic fibrosis show an increased level of antioxidants and detoxification enzymes in the areas of epithelial regeneration (in fibrotic lesions the antioxidants are low). It seems that a global alteration of redox balance is more important than the trend of the single oxidant or antioxidant factors. This redox deregulation causes an activation of the epithelial TGF beta and proteases and an increase in extracellular matrix production that may ultimately contribute to the development of the end-stage fibrosis [15–17]. Despite the clear involvement of oxidant stress in chronic lung diseases, current potential treatments, such as the administration of several antioxidants, failed to protect the respiratory system against COPD and IPF. In the next paragraphs, we will briefly take into consideration the cross-talk between inflammatory and oxidative stress mediators and the possible role of the indirect stimulator of antioxidant and anti-inflammatory mediators, such as exercise, caloric reduction, or ozone therapy, in the prevention and treatment of chronic lung diseases.

1.1 Oxygen and oxidant molecules in the respiratory system: the necessary evil?

1.1.1 Source of oxidative stress in the chronic pulmonary diseases

In the lung, the main sources of oxidants are high oxygen tension, environmental factors, such as cigarette smoke and airborne particulates, and infections. If the concentrations of oxidants remain high for a long time in the pulmonary tissue, chronic oxidative stress could begin. As far as the oxygen tension in the atmosphere, there is no doubt that it represents both challenges and opportunities for life. It is universally accepted that the energy generates through O_2 electron transport is a crucial mechanism for the evolution of complex multicellular organisms. However, about 2–3% of oxygen used by mitochondria, via the complex I, II, and III, during the process of oxidative phosphorylation will leak from the respiratory chain to form superoxide anion $O_2^{\bullet -}$. On the basis of recent experimental findings, the mitochondrial source of superoxide anions is much less quantitatively relevant with respect to the previous estimation. Other mitochondrial enzymes, such as dihydrolipoamide, dehydrogenase, and monoamine oxidase, are able to produce ROS [18, 19]. NAD(P)H oxidases present in cell membranes of fibroblasts, endothelial, and vascular smooth muscle cells, and particularly phagocytes, produce superoxide as a basic defensive process [20]. Other enzymes, such as nitric oxide synthase (NOS), xanthine oxidase, cytochrome P450, lipoxygenases, and even heme oxygenases (HOs), during the abnormal situation as inflammation, may be implicated in superoxide production. The reduction of superoxide anion, discovered by McCord and Fridovich in 1968, is performed by mitochondrial (SOD Mn), cytosolic (SOD Cu/Zn), and extracellular superoxide dismutases (SODs), that catalyze the dismutation to hydrogen peroxide as follows:

$$2O_2^{\bullet -} + 2\,H \rightarrow H_2O_2 + O_2 \tag{1}$$

Hydrogen peroxide (H_2O_2) is not a radical molecule because it has paired electrons but it has been included among the reactive oxygen species (ROS) for its oxidant property [21]. As it is a unionized molecule, in the presence of an extracellular-cytosolic gradient, it passes throughout the cell membrane but the intracellular concentration is only about 1/10 of the extracellular one. The hydrogen peroxide has a half-life of about 1–2 seconds in plasma but less than 1 second in blood. The concentration of H_2O_2 in plasma is about 2.5 micromoles; it appears ubiquitous and it has been detected in urine and in exhaled air of patients affected by chronic lung diseases [22]. Hydrogen peroxide, due to its relatively mild reaction, can serve as the main component of intracellular signal transduction. However, when the concentration of ROS exceeds the homeostatic level, it can generate another most potent free radical, the Hydroxyl radical (OH^-) via the Fenton-Jackson reaction.

$$H_2O_2 + Fe^{++} \rightarrow OH\bullet + OH^- + Fe^{+++} \tag{2}$$

Hydroxyl radicals can cause covalent cross-linking of enzymes or propagates deleterious free radical reactions in a large series of molecules, such as DNA, proteins, and lipids. As far as the lipid interaction, hydroxyl radicals can eliminate a hydrogen atom from PUFA resulting in the formation of lipid radicals, which can interact further with oxygen to generate the lipid peroxyl radical. A lipid peroxidation occurs if the lipid peroxyl radical is not completely reduced by antioxidants. Lipid peroxidation products (LOP) are generally stable, can diffuse within or even escape from the cell and attack the target far from the site of the original free radical event. To avoid all these undesirable consequences, glutathione peroxidase and catalase are enzymes that play a pivotal role in preventing hydroxyl radical formation because they can definitively convert the hydrogen peroxide to water [23, 24]. For many years ROS generation was regarded as a damaging side effect of aerobic metabolism and only toxic effects were attributed to ROS. Recently, in contrast, it was recognized that ROS are part of cellular redox homeostasis and the complete elimination of these molecules would disrupt rather than extend the normal function of the organism [25]. In fact, the physiological production of ROS regulates many redox-dependent signaling processes that control proliferation, migration, differentiation, or cell survival by inducing specific and reversible posttranslational modifications on redox-sensitive proteins. Whether ROS will act as damaging, protective, or signaling factors depends on the delicate equilibrium between ROS production and scavenging at the proper time and side [26, 27]. In conclusion, oxygen and its derived species could be considered essential consequences and drivers for evolution and survival over Earth's history, but an abnormal presence of ROS is associated with the development of chronic oxidative stress diseases. Only the correct understanding of the physiologic O_2 concentration and ROS threshold may provide novel insight into innovative strategies for treating a large number of degenerative pathologies, including chronic lung diseases [28, 29].

1.2 The antioxidant system in lung

In the human body, all organs contain a large number of antioxidant molecules to prevent inappropriate ROS production or unwanted action of ROS in cells. In the lung, the respiratory tract epithelial lining fluid (RTLF) represents a physical barrier between the external environment and the underlying respiratory tract epithelial cell layer. In this lining fluid, most of the inhaled toxicants are cleared from the lungs also thanks to the mucociliary action. A large amount of nonenzymatic low molecular weight antioxidant scavengers are contained in the RTLF, and these substances are able to directly counteract and detoxify the inhaled oxidants and thereby prevent the direct contact of inhaled toxicants with the underlying epithelium. The major antioxidant molecules in the RTLF are glutathione (GSH), ascorbic acid, uric acid, and vitamin E. Additionally, airway epithelial cells secrete certain antioxidant proteins into RTLF, which also function as antioxidant scavengers. Other ROS metabolizing enzymes in the lung tissue are superoxide dismutases (SODs), catalases, peroxiredoxins, glutathione peroxidase, thiol reductases that reverse the modifications of cysteine after oxidation, thioredoxin (TRX, glutaredoxin (GRX), phase-2 detoxifying enzymes (e.g., glutathione S-transferases (GST), metal-binding proteins (transferrin and lactoferrin). Under physiological conditions, there is an equilibrium between intracellular ROS and endogenous antioxidants. The mechanism of antioxidant defense is complex and compartmentalized allowing independent regulation of cytoplasmic, mitochondrial, and nuclear levels of ROS. Superoxide dismutase (SOD) are intracellular enzymes and represent the first line of protection against ROS. They catalyze the dismutation of $O_2^{\cdot-}$ in H_2O_2. There are three types of superoxide dismutase with different localization: Copper-zinc SOD (SOD1 and cytoplasmic), manganese SOD (SOD2 and mitochondrial), and zinc SOD (SOD3 and extracellular SOD) [30]. The action of SOD must be coupled with that of enzymes that degrade $H_2O_{2,}$ such as catalase or glutathione peroxidase, in order to avoid increasing concentration of $H_2O_{2,}$ which in presence of iron induces the formation of OH radical by the Fenton reaction. Catalase, various peroxidases, including glutathione peroxidase and glutathione S transferase, can convert H_2O_2 to H_20. Other enzymes act with direct detoxification of ROS. Glutathione is a tripeptide formed by glutamic acid, cysteine, and glycine. It is represented in a simplified way by GSH (reduced form) or glutathione disulfide (GSSG) oxidized form. GSH can remove the ROS either by a direct chemical reaction or via peroxide reduction as the co-factor of GSH peroxidase inducing a cycle between the reduced and the oxidized form of glutathione [31]. The reduced form is maintained by GSH reductase. The thioredoxin/thioredoxin reductase system facilitates the reduction of other proteins by the formation of disulfide bridges between cysteine residues. Recent experimental findings suggest that each component of this defense system has a specific function and these components are not interchangeable. This is an important concept because it means that an increased amount of one does not compensate for a deficiency of other [32].

2. Reactive oxygen species and inflammatory mediators: the dangerous crosslink

Reactive oxygen species are generally considered as proinflammatory molecules and most studies assumed that increased production of any ROS will result in cellular and organ dysfunction because of a rapid transition from one reactive oxygen intermediate to another. On the contrary, a large series of experimental findings suggest that different ROS may have distinct effects on individual cell activation. ROS have far-reaching effects on the respiratory tract and the parenchyma and increase the inflammatory response ROS activates nuclear factor-kappa beta (NF-kB), which turns on multiple inflammatory genes, resulting in enhancing the inflammatory response, oxidative stress leads to the activation of histone acetyltransferase activity, which opens the chromatin structure and is with increased transcription of multiple inflammatory genes [33]. Oxidative stress can also affect the function of antiproteases, such as alpha1-antitrypsin and secretory leukoprotease inhibitor, and thus accelerates the breakdown of elastin in the pulmonary parenchyma [34]. Patients with COPD, especially in severe disease and during exacerbations, have evidence of systemic inflammation measured either as elevated circulating cytokine, chemokine, and acute phase protein levels or as anomalies in circulating cells [35]. Persistent inflammation is associated with worse clinical results. Smoking itself can produce systemic inflammation (e.g., increased total number of leukocytes), but in patients with COPD, the degree of systemic inflammation is a sum of local and systemic inflammation [35]. The systemic inflammation in patients with COPD may contribute to its systemic manifestations and could aggravate extra-pulmonary manifestations. A clinical study measured a series of plasma inflammatory markers (C-reactive protein, IL-6, CXCL8, fibrinogen, TNF-a, and leukocytes) in COPD patients and the results showed that 70% of patients with COPD had some components of systemic inflammation, and 16% had persistent inflammation [36]. Patients with persistent systemic inflammation had increased mortality and more frequent exacerbations. Systemic inflammation appears to relate to an accelerated decrease in lung function and is increased further during exacerbations. Moreover, oxidative stress drives accelerated aging through activation of phosphoinositide3-kinase (PI3K) and reduction in sirtuin-1 levels, which leads to cellular senescence and release of inflammatory proteins, which further increase oxidative stress [37, 38]. In patients affected by pulmonary fibrosis, evidence strongly suggest the important contributions of the oxidants and inflammatory mediators to the pathogenesis of IPF [39–41]. A large series of studies suggested an important contribution of ROS and RNS to the proteases/anti-proteases imbalance through both activation of matrix metalloproteinases (MMP) and inactivation of protease inhibitors. It has also been suggested that ROS can both activate and inactivate MMPs, it depends on the local amount and distribution of ROS [42]. The oxidative stress and the proteolytic activation are mutually reinforcing processes that leading to tissue injury and lung fibrosis [43–45]. It is therefore clear that the prevention or the therapy of chronic lung disease must take into account the interdependence between oxidant and inflammatory systems because they are closely related.

2.1 Redox modulation in the prevention and treatment of chronic lung diseases

A large series of studies suggested a connection between impairment of the antioxidant defense system controlled by nuclear factor erythroid 2-related factor (Nrf2) and chronic lung disease development and progression [46]. Nrf2 plays a central role in controlling redox homeostasis and it is involved in the induction of several enzymes implicated in the antioxidant defense such as heme oxygenase (HO)-1, glutamate-cysteine ligase modifier subunit (GCLM) and glutamate-cysteine ligase catalytic subunit (GCLC) but also in the anti-inflammatory regulation, such as transforming growth factor (TGF)—beta and nuclear factor kappa (NF-k)beta. In homeostatic conditions, Nrf2 is tied in the cytosol to its repressor Kelch-like ECH-associated protein 1 (Keap1) and cullin3-dependent E3 ubiquitin ligase. Two domains of Keap1 protein contain key reactive cysteine residues and the modification of these residues causes the disruption of the Nrf2-keap1 complex and the consequent activation of Nrf2. The free Nrf2 migrates in the nucleus and binds, as a heterodimer with small Maf proteins, to the antioxidant response element (ARE) in the upstream promoter region of antioxidant and phase II detoxifying enzymes genes and initiates transcription and expressions of these proteins [47]. Stimulation of Nrf2 is upregulated during short exposure to a low-intensity stressor. It has been reported a clear association between reduced Nrf2 signal pathway and the development and progression of COPD [48, 49]. For example, in an experimental setting with genetically null Nrf2 mice exposed to chronic cigarette smoke, increased alveolar destruction and inflammation were observed compared to mice with normal expression of Nrf2. In fact, in Nrf2 −/− mice, the level of histone deacetylases 6 (HDAC6) was increased and this data suggests that Nrf2 could counteract the increased expression of HDAC6 by oxidative and proteolytic stress [50–52]. In fibroblast of IPF patients, it has been reported a decrease in expression and nuclear localization of Nrf2, with a reduction of related genes (HO-1, NQO1, and epoxide hydrolase). Moreover, an increase in alpha-smooth muscle actin and collagen was observed suggesting a conversion from fibroblast to myofibroblast phenotype [53]. Nrf2 is a key factor in aging and a predisposing factor for chronic lung diseases. In fact, the capacity to respond to oxidative stress through the Nrf2 activation decreases with aging. This phenomenon is probably due to many factors like the reduction of positive regulators of Nrf2 (PI3K, P62, CPB, and BRCA1) and the increase in the Nrf2 suppressors (Keap1, Bach1, and cMyc) with age [54, 55]. All together these experimental data strongly suggest that compounds able to activate Nrf2 or stabilize Keap1/DJ-1/Maf proteins could exert an important role in the prevention and therapy of chronic lung diseases, especially in situations where the endogenous antioxidant system is weakened (advanced COPD) or is less adaptative/compensatory or decreased (e.g., in IPF or aging lung) [56]. In fact, on the basis of the previously exposed mechanisms involved in the pathogenesis of COPD and IPF, it is possible to speculate that decreasing ROS overproduction or increasing antioxidants in the lung could lead to reduced pulmonary damage and a consequent systemic polymorbidity [57]. The most important Nrf2 activators are dietary and synthetic products containing sulforaphane, curcumin, and caffeic-acid phenethyl ester, which are able to induce ARE-regulated gene expression and could be also useful in chemoprevention [58, 59]. Chalcones have been tested as both redox modulators and anti-inflammatory molecules because they simultaneously inhibit the NF-kB pathway and activate Nrf2/ARE pathways with consequent induction of the expression of phase II detoxifying enzymes. Thiol compounds (N acetylcysteine and carbocysteine) have been tried in the prevention and treatment of COPD and IPF [60–62]. Unfortunately, the clinical trials of these compounds and other molecular antioxidants, such as vitamin C and E or superoxide dismutase (SOD), showed contrasting results in the treatment of chronic lung diseases [63]. It has been suggested that the intracellular bioavailability and the heterogeneity of the patients could be the main issues in the efficacy of these therapeutic agents. Further studies on the combination of oral administration of several redox modulators were performed, but it has been reported that antioxidants, if exogenously oversupplied, can cause antioxidant stress because, as reported above, ROS are signaling molecules during the homeostatic cellular process and the full suppression of this molecules causes damage. Thus, the amount of ROS in the cellular environment can become protective or harmful depending on a fine and complex redox modulation at the proper side and time [64–67].

3. The future challenge in the prevention and treatment of chronic lung diseases

To better achieve the therapeutical effects in COPD and IPF patients, clinical trials used redox modulators at high dosages for months. The aim of this protocol is to maintain a continuous high level of Nrf2 activation in the cells. Unfortunately, prolonged and excessive Nrf2 activation may have detrimental effects, as suggested by the relationship between Nrf2 overexpression and cancer promotion or cancer cell protection from chemotherapy in lung cancer [68]. In fact, in chronic lung diseases, the association with aging lungs and COPD or IPF alterations could produce cancer development in the presence of high levels of Nrf2 activators. Other protocols of antioxidant's administration reported that inhaled N acetylcysteine(NAC) monotherapy improves redox balance in IPF patients [69]. However, the only valid mechanism able to activate the Nrf2 pathway and the intracellular antioxidants generation is an "on and off" mechanism that could restore the oxidant/antioxidant balance and preserve the cells from the detrimental consequences of prolonged Nrf2 stimulation. For this purpose, particularly in the prevention of chronic lung diseases, it has been shown that regular exercise is one of the most successful interventions to prevent or delay chronic diseases. Aerobic exercise induces a number of biochemical signaling messages that result in changes in cell physiology, many of which are mediated by redox mechanisms. Calculated and transient redox stress caused by acute exercise increases Nrf2 activation in young animals and humans, thus improving cellular resistance to subsequent redox stressors [70, 71]. A significant increase in gene expression of Nrf2 and target SOD2 were shown in skeletal muscle of young fit males following an acute bout of cycling exercise lasting 90 min in normoxic recovery conditions. Mode, intensity, and duration of exercise could each impact the rate and amplitude of Nrf2 cycling *in vivo* [72]. Caloric restriction is also an endogenous Nrf2 activator, studies have shown that nutritional components may modulate the Nrf2-Keap1 system, so it may be of fundamental importance to demonstrate the beneficial effects of this system in various chronic diseases [73–75]. Other studies reported that during ozone therapy (major ozonated autohemotherapy) a transient and calculating stimulation of Nrf2 occurs [76]. In fact, ozone therapy is able to activate Nrf2 for the production of antioxidants and phase II enzymes and at the same time, it is unable to prolong this activity after about 40–60 min. This mechanism of transient Nrf2 stimulation could be very useful in the lung cells undergoing oxidative stress in chronic lung diseases for restoring a normal redox system without the risk of cellular proliferation subsequent to a continuous prolonged Nrf2 stimulation.

4. Conclusions

Experimental and clinical evidences strongly suggest that maintenance of a proper redox balance through regulation of the Nrf2/ARE pathway could be critical to the integrity and function of intracellular components in the respiratory system. It is clear that both development and progression of chronic lung diseases may be caused by an impairment of Nrf2 activation or excessive Nrf2 stimulation. The compounds that boost Nrf2 activity showed promising results but bioavailability and potential risks and benefits remain to be proved. An endogenous redox modulator like ozone, if correctly used, seems to be able to elicit a transient and calculate oxidative stress and enzymatic production of antioxidants and other detoxifying enzymes [77]. Exercise and caloric restriction are also possible indirect ways to stimulate the Nrf2/ARE pathway [78, 79]. Future studies are needed to decipher the intracellular signals involved in the activation of Nrf2/ARE- mediate gene transcription by oxidative stress and to evaluate their role in the development of chronic lung diseases.

Author details

Emma Borrelli
Department of Medical Biotechnologies, University of Siena, Siena, Italy

*Address all correspondence to: borrelli@unisi.it

References

[1] Holguin F. Oxidative stress in airway diseases. Annals of the American Thoracic Society. 2013;**10**:S150-SS15

[2] Kirkham PA, Barnes PJ. Oxidative stress in COPD. Chest. 2013;**1**:266-273

[3] Van der Vliet A, Janssen-Heininger YMW, Anathy V. Oxidative stress in chronic lung disease: From mithocondrial dysfunction to dysregulated redox signaling. Molecular Aspects of Medicine. 2018;**63**:59-69

[4] Fischer BM, Pavlisko E, Voynow JA. Pathogenic triad in COPD: Oxidative stress, protease-antiprotease imbalance and inflammation. International Journal of Chronic Obstructive Pulmonary Disease. 2011;**6**:413-421

[5] Divo M, Cote C, de Torres JP, et al. Comorbidities and risk of mortality in patients with chronic obstructive pulmonary disease. American Journal of Respiratory and Critical Care Medicine. 2012;**186**:155-161

[6] Fabbri LM, Rabe KF. From COPD to chronic systemic inflammatory syndrome? Lancet. 2007;**370**:797-799

[7] Antus B, Kardos Z. Oxidative stress in COPD: Molecular background and clinical monitoring. Current Medicinal Chemistry. 2015;**22**(5):627-650. DOI: 10.2174/0929867322205150112104411

[8] Choudhury G, MacNee W. Role of inflammation and oxidative stress in the pathology of ageing in COPD: Potential therapeutic interventions. COPD. 2017;**14**:122-135

[9] Hoshino Y, Mishima M. Redox-based therapeutics for lung diseases. Antioxidants & Redox Signaling. 2008;**10**:701-704

[10] Wang C, Zhou J, Wang J, Li S, Fukunaga A, Yodoi J, et al. Progress in the mechanism and targeted drug therapy for COPD. Signal Transduction and Targeted Therapy. 2020;**5**:248

[11] Mastruzzo C, Crimi N, Vancheri C. Role of oxidative stress in pulmonary fibrosis. Monaldi Archives for Chest Disease. 2002;**57**:173-176

[12] Kinnula VL, Fattman CL, Tan RJ, et al. Oxidative stress in pulmonary fibrosis: A possible role for redox modulatory therapy. American Journal of Respiratory and Critical Care Medicine. 2005;**172**:417-422

[13] Montuschi P, Ciabattoni G, Paredi P, Pantelidis P, du Bois RM, Kharitonov SA, et al. 8-isoprostane as a biomarker of oxidative stress in interstitial lung diseases. American Journal of Respiratory and Critical Care Medicine. 1998;**158**:1524-1527

[14] Rahman I, Macnee W. Oxidative stress and regulation of glutathione in lung inflammation. The European Respiratory Journal. 2000;**16**:534-554

[15] Cheresh P, Kim SJ, Tulasiram S, Kamp DW. Oxidative stress and fibrosis. Biochimica et Biophysica Acta. 2013;**832**:1028-1040

[16] Lee CM, Park JW, Cho WK, Zhou Y, Han B, Yoon PO. Modifiers of TGF-beta 1 effector function as novel therapeutic targets of pulmonary fibrosis. The Korean Journal of Internal Medicine. 2014;**29**:281-290

[17] Fois AG, Paliogiannis P, Sotgia S, Mangoni AA, Zinellu E, Pirina P, et al.

Evaluation of oxidative stress biomarkers in idiopathic pulmonary fibrosis and therapeutic applications: A systematic review. Respiratory Research. 2018;**19**(1):51. DOI: 10.1186/s12931-018-0754-7

[18] Sies HR, Jones GP. Reactive oxygen species (ROS) as pleiotropic physiological signalling agents. Nature Reviews. Molecular Cell Biology. 2020;**21**:363-383. DOI: 10.1038/s41580-020-0230-3

[19] Ostaszewska-Bugajska M, Podgórska A, Szal B. Markers for mitochondrial ROS status. Methods in Molecular Biology. 2022;**2363**:199-213. DOI: 10.1007/978-1-0716-1653-615

[20] Begum R, Thota S, Abdulkadir A, Kaur G, Bagam P, Batra S. NADPH oxidase family proteins: Signaling dynamics to disease management. Cellular & Molecular Immunology. 2022;**19**(6):660-686. DOI: 10.1038/s41423-022-00858-1

[21] Winterbourn CC. The biological chemistry of hydrogen peroxide. Methods in Enzymology. 2013;**528**:3-25. DOI: 10.1016/B978-0-12-405881-1.00001-X

[22] Murata K, Fujimoto K, Kitaguchi Y, Horiuchi T, Kubo K, Honda T. Hydrogen peroxide content and pH of expired breath condensate from patients with asthma and COPD. COPD. 2014;**11**(1):817. DOI: 10.3109/15412555.2013.830094

[23] Halliwell B. Free radicals, proteins and DNA: Oxidative damage versus redox regulation. Biochemical Society Transactions. 1996;**24**:1023-1027

[24] Mumby S, Adcock IM. Recent evidence from omic analysis for redox signalling and mitochondrial oxidative stress in COPD. Journal of Inflammation. 2022;**19**(1):10. DOI: 10.1186/s12950-022-00308-9

[25] Sadiq IZ. Free radicals and oxidative stress: Signaling mechanisms, redox basis for human diseases, and cell cycle regulation. Current Molecular Medicine. 2021;**23**:13-35. DOI: 10.2174/1566524022666211222161637

[26] Valko M, Leibfritz D, Moncol J, Cronin MT, Mazur M, Telser J. Free radicals and antioxidants in normal physiological functions and human disease. The International Journal of Biochemistry & Cell Biology. 2007;**39**:44-84

[27] Thannickal VJ, Fanburg BL. Reactive oxygen species in cell signaling. Am J Physiol Cell Mol Physiol. 2000;**279**:L1005-L1028

[28] Kurutas EB. The importance of antioxidants which play the role in cellular response against oxidative/nitrosative stress: Current state. Nutrition Journal. 2015;**15**:71-80

[29] Rahman I, Adcock IM. Oxidative stress and redox regulation of lung inflammation in COPD. Eur Resp J. 2006;**28**:219-242

[30] Kinnula VL, Crapo JD. Superoxide dismutases in lung and human lung diseases. American Journal of Respiratory and Critical Care Medicine. 2003;**167**:1600-1619

[31] Watson WH, Ritzenthaler JD, Peyrani P, Wiemken TL, Furmanek S, Reyes Vega AM, et al. Plasma cysteine/cystine and glutathione/glutathione disulfide redox potentials in HIV and COPD patients. Free Radical Biology & Medicine. 2019;**143**:55-61. DOI: 10.1016/j.freeradbiomed.2019.07.031

[32] Halliwell B. Cell culture, oxidative stress, and antioxidants: Avoiding

pitfalls. Biomed J. 2014;**37**:99-105. DOI: 10.4103/2319-4170.128725

[33] Morgan MJ, Liu ZG. Crosstalk of reactive oxygen species and NF-kappa B signaling. Cell Research. 2006;**21**:103-115

[34] Di Stefano A, Caramori G, Oates T, Capelli A, Lusuardi M, Gnemmi I, et al. Increased expression of nuclear factor kappa B in bronchial biopsies from smokers and patients with COPD. The European Respiratory Journal. 2002;**3**:556-563

[35] Caramori G, Adkock IM, Di Stefano A, Chung KF. Cytokine inhibition in the treatment of COPD. Int J Chronic Obstr Pulm Dis. 2014;**9**:397-412

[36] Celli BR, Locantore N, Yates J, Tal-Singer R, Miller BE, Bakke P, et al. ECLIPSE investigators inflammatory biomarkers improve clinical prediction of mortality in chronic obstructive pulmonary disease. American Journal of Respiratory and Critical Care Medicine. 2012;**15**(185):1065-1072. DOI: 10.1164/rccm.201110-1792OC

[37] Baker J, Vuppusetty C, Fenwick CT, P, Donnelly L, Ito K, Barnes PJ. Oxidative stress dependent micro RNA-34a activation via PI3Kalpha reduces the expression of sirtuin-1 and sirtuin-6 in epithelial cells. Scientific Reports. 2016;**6**:35871

[38] Vanfleteren LE, Spruit MA, Groenen M, Gaffron S, van Empel VP, Bruijnzeel PL, et al. Clusters of comorbidities based on validated objective measurements and systemic inflammation in patients with chronic obstructive pulmonary disease. American Journal of Respiratory and Critical Care Medicine. 2013;**187**:728-735

[39] Kinnula VL, Myllarniemi M. Oxidant-antioxidant imbalance as a potential contributor to the progression of human pulmonary fibrosis. Antioxidants & Redox Signaling. 2008;**10**:727-738

[40] Cantin AM, North SL, Fells GA, Hubbard RC. Crystal RG oxidant-mediated epithelial cell injury in idiopathic pulmonary fibrosis. The Journal of Clinical Investigation. 1987;**79**:1665-1673

[41] Cui Y, Robertson J, Maharaj S, Waldhauser L, Niu J, Wang J, et al. Oxidative stress contributes to the induction and persistence of TGF-β1 induced pulmonary fibrosis. The International Journal of Biochemistry & Cell Biology. 2011;**43**:1122-1133

[42] Pardo A, Cabrera S, Maldonado M, Selman M. Role of matrix metalloproteinases in the pathogenesis of idiopathic pulmonary fibrosis. Respiratory Research. 2016;**17**:23

[43] Bringardner BD, Baran CP, Eubank TD, Marsh CB. The role of inflammation in the pathogenesis of idiopathic pulmonary fibrosis. Antioxidants & Redox Signaling. 2008;**10**:287-301

[44] Otoupalova E, Smith S, Cheng G, Thannickal V. Oxidative stress in pulmonary fibrosis. Comprehensive Physiology. 2020;**10**:509-547

[45] Gadek JE, Fells GA, Zimmerman RL, Keogh BA, Crystal RG. Role of proteolytic and oxidative products of the neutrophil in determining the specificity of the pulmonary lesion in fibrotic and destructive lung disease. Chest. 1983;**83**:59S. DOI: 10.1378/chest.83.5_supplement.59s

[46] Nguyen T, Nioi P. Pickett CB the Nrf2-antioxidant response element signaling pathway and its activation by

oxidative stress. The Journal of Biological Chemistry. 2009;**284**:13291-13295

[47] Boutten A, Goven D, Artaud-Macari E, Boczkowski J, Bonay M. NRF2 targeting: A promising therapeutic strategy in chronic obstructive pulmonary disease. Trends in Molecular Medicine. 2011;**17**:363-371

[48] Cui W, Zhang Z, Zhang P, Qu J, Zheng C, Mo X, et al. Gao J Nrf2 attenuates infammatory response in COPD/emphysema: Crosstalk with Wnt3a/beta-catenin and AMPK pathways. Journal of Cellular and Molecular Medicine. 2018;**22**:3514-3525

[49] Liu Q, Gao Y, Ci X. Role of Nrf2 and its activators in respiratory diseases. Oxidative Medicine and Cellular Longevity. 2019;**2019**:7090534

[50] Artaud-Macari E, Goven D, Brayer S, Hamimi A, Besnard V, Marchal-Somme J, et al. Nuclear factor erythroid 2-related factor 2 nuclear translocation induces myofibroblastic dedifferentiation in idiopathic pulmonary fibrosis. Antioxidants and Redox Signaling. 2013;**18**:66-79

[51] Iizuka T, Ishii Y, Itoh K, Kiwamoto T, Kimura T, Matsuno Y, et al. Nrf2-deficient mice are highly susceptible to cigarette smoke-induced emphysema. Genes to Cells. 2005;**10**:1113-1125

[52] Rangasamy T, Cho CY, Thimmulappa RK, Zhen L, Srisuma SS, Kensler TW, et al. Genetic ablation of Nrf2 enhances susceptibility to cigarette smoke-induced emphysema in mice. The Journal of Clinical Investigation. 2004;**114**:1248-1259. DOI: 10.1172/JCI21146

[53] Hii Y, Itoh K, Morishima Y, Kimura T, Kiwamoto T, Iizuka T, et al. Transcription factor Nrf2 plays a pivotal role in protection against elastase-induced pulmonary inflammation and emphysema. Journal of Immunology. 2005;**175**:6968-6975

[54] Fratta Pasini AM, Stranieri C, Ferrari M, Garbin U, Cazzoletti L, Mozzini C, et al. Oxidative stress and Nrf2 expression in peripheral blood mononuclear cells derived from COPD patients: An observational longitudinal study. Respiratory Research. 2020;**21**:37-44

[55] Itoh K, Tong KI, Yamamoto M. Molecular mechanism activating Nrf2-Keap1 pathway in regulation of adaptive response to electrophiles. Free Radical Biology & Medicine. 2004;**36**:1208-1e13

[56] Baird L, Dinkova-Kostova LT. The cytoprotective role of the Keap1-Nrf2 pathway. Archives of Toxicology. 2011;**85**:241-272

[57] Leinonen HM, Kansanen E, Polonen P, Heinaniemi M, Levonen AL. Dysregulation of the Keap1-Nrf2 pathway in cancer. Biochemical Society Transactions. 2015;**43**:645-649

[58] Sadowska AM, Manuel-Y-Keenoy B, De Backer WA. Antioxidant and anti-inflammatory efficacy of NAC in the treatment of COPD: Discordant in vitro and in vivo dose-effects: A review. Pulmonary Pharmacology & Therapeutics. 2007;**20**:9-22

[59] Zheng JP, Wen FQ, Bai CX, Wan HY, Kang J, Chen P, et al. Twice daily Nacetylcysteine 600 mg for exacerbations of chronic obstructive pulmonary disease (PANTHEON): A randomised, doubleblind placebo-controlled trial. The Lancet Respiratory Medicine. 2014;**2**:187-194

[60] Fowdar K, Chen H, He Z, Zhang J, Zhong X, Zhang J, et al. The effect of

N-acetylcysteine on exacerbations of chronic obstructive pulmonary disease: A meta-analysis and systematic review. Heart & Lung. 2017;**46**:120-128

[61] Demedts M, Behr J, Buhl R, Costabel U, Dekhuijzen R, Jansen HM, et al. High-dose acetylcysteine in idiopathic pulmonary fbrosis. The New England Journal of Medicine. 2005;**353**:2229-2242

[62] Homma S, Azuma A, Taniguchi H, Ogura T, Mochiduki Y, Sugiyama Y, et al. Efficacy of inhaled N-acetylcysteine monotherapy in patients with early stage idiopathic pulmonary fibrosis. Respirology. 2012;**17**:467-477

[63] Magesh S, Chen Y, Hu L. Small molecule modulators of Keap1-Nrf2-ARE pathway as potential preventive and therapeutic agents. Medicinal Research Reviews. 2012;**32**:687-726

[64] Guerrero-Beltrán CE, Calderón-Oliver M, Pedraza-Chaverri J, Chirino YI. Protective effect of sulforaphane against oxidative stress: Recent advances. Experimental and Toxicologic Pathology. 2012;**64**:503-508

[65] Cook-Mills JM, Abdala-Valencia H, Hartert T. Two faces of vitamin E in the lung. American Journal of Respiratory and Critical Care Medicine. 2013;**188**(3):279-284

[66] Bocci V, Valacchi G. Nrf2 activation as target to implement therapeutic treatments. Frontiers in Chemistry. 2015;**3**:4. DOI: 10.3389/fchem.2015.00004

[67] Rahman I. MacNee W antioxidant pharmacological therapies for COPD. Current Opinion in Pharmacology. 2012;**12**:256-265

[68] Pouremamali F, Pouremamali A, Dadashpour M, Soozangar N, Jeddi F. An update of Nrf2 activators and inhibitors in cancer prevention/promotion. Cell Communication and Signaling: CCS. 2022;**20**:100

[69] Muramatsu Y, Sugino K, Ishida F, Tatebe J, Morita T, Homma S. Effect of inhaled N-acetylcysteine monotherapy on lung function and redox balance in idiopathic pulmonary fibrosis. Respiratory Investigation. 2016;**54**(3):1708. DOI: 10.1016/j.resinv.2015.11.004

[70] Muthusamy VR, Kannan S, Sadhaasivam K, Gounder SS, Davidson CJ, Boeheme C, et al. Acute exercise stress activates Nrf2/ARE signaling and promotes antioxidant mechanisms in the myocardium. Free Radical Biology & Medicine. 2012;**52**:366-376

[71] Merry TL, Ristow M. Nuclear factor erythroid-derived 2-like 2 (NFE2L2, Nrf2) mediates exercise-induced mitochondrial biogenesis and the anti-oxidant response in mice. The Journal of Physiology. 2016;**594**:5195-5207

[72] Cases N, Sureda A, Maestre I, Tauler P, Aguiló A, Córdova A, et al. Response of antioxidant defences to oxidative stress induced by prolonged exercise: Antioxidant enzyme gene expression in lymphocytes. European Journal of Applied Physiology. 2006;**98**(3):263-269. DOI: 10.1007/s00421-006-0273-y

[73] Schols AM. Nutrition as a metabolic modulator in COPD. Chest. 2013;**144**:1340-1345

[74] Nakagawa F, Morino K, Ugi S, Ishikado A, Kondo K, Sato D, et al. 4-Hydroxy hexenal derived from dietary n-3 polyunsaturated fatty acids induces anti-oxidative enzyme heme oxygenase-1 in multiple organs. Biochemical and

Biophysical Research Communications. 2014;**443**:991-996

[75] Martín-Montalvo A, Villalba JM, Navas P, de Cabo R. NRF2, cancer and calorie restriction. Oncogene. 2011;**30**:505-520

[76] Bocci V, Borrelli E, Travagli V, Zanardi I. The ozone paradox. Ozone is a strong oxidant as well as medical drug. Medicinal Research Reviews. 2009;**29**:646-682

[77] Borrelli E. The problem of chronic oxidative stress and the systemic indirect ozone therapy As a possible solution. Highlights on Medicine and Medical Research. 2021;**4**:135-140. DOI: 10.9734/bpi/hmmr/v4/1739F

[78] Dimauro I, Paronetto MP, Caporossi D. Exercise, redox homeostasis and the epigenetic landscape. Redox Biology. 2020;**35**:101477

[79] Milder JB, Liang LP, Patel M. Acute oxidative stress and systemic Nrf2 activation by the ketogenic diet. Neurobiology of Disease. 2010;**40**:238-244

Chapter 5

Cognitive Impairment in Diabetes Mellitus and Its Management by Transcription Factor Nrf2-Mediated Antioxidant Defense System

Xian Zhou, Gerald Münch and Dennis Chang

Abstract

Diabetes mellitus has been an epidemic in the twenty-first century and an approximately 50% risk of diabetes predisposed to cognitive decline leading to dementia in humans. There is an urgent need to understand the pathophysiology and identify molecular targets of cognitive impairment in diabetes mellitus that might lead to improved therapy. Mounting evidence indicates that nuclear factor erythroid 2-related factor 2 (Nrf2) and its regulated downstream antioxidant genes are emerging therapeutic targets. In this chapter, we introduce cognitive dysfunction in diabetes mellitus and its hallmarks, particularly its pathological mechanisms related to oxidative stress in the brain, then justify the role of the transcription factor Nrf2-mediated antioxidant defense system in attenuating cognitive decline in diabetes mellitus. Studies on Nrf2 inducers sourced from natural products (i.e., sulforaphane, astaxanthin, resveratrol, quercetin) that have shown potent cognitive improvement in diabetic models are discussed. These studies have demonstrated that Nrf2 inducers drive the antioxidant and anti-inflammatory responses in the hippocampus region and effectively improve the spatial and memory function in diabetic rats/mice. However, evidence from large and well-designed clinical trials is warranted to support Nrf2 inducers as promising therapeutic agents in the management of cognitive impairment in diabetes mellitus.

Keywords: cognitive impairment in diabetes mellitus, Nrf2, natural compounds, oxidative stress, NLRP3 inflammasome, neuroinflammation

1. Introduction

Diabetes mellitus (DM) is a chronic disease characterized by excessively high levels of blood glucose either resulting from insulin resistance or inadequate insulin secretion [1]. High blood glucose levels (hyperglycemia) in DM can cause serious damage to the heart, blood vessels, eyes, kidneys and nerves, and thus DM complications can significantly impact the quality of life and reduce life expectancy [2]. DM neuropathy is one of the major DM complications affecting 50% of DM sufferers which describes a type of nerve damage throughout the body induced by hyperglycemia [3, 4]. In the central nervous system, both type 1 (T1DM) and type 2 DM (T2DM) can cause nerve damage and result in various degrees of cognitive decline. Mounting epidemiological studies showed that the risk of cognitive decline in middle-aged diabetic patients was 2–3 times higher compared with healthy peers [5, 6]. Moreover, the prevalence of cognitive impairment is predicted to increase dramatically in the future with a longer survival time for diabetic patients [7]. Cognitive decline among children with T1DM is mainly manifested as subtle changes in cognitive development, whereas adults present subtle decrements in cognitive performance compared to age-matched controls [8]. The severity of cognitive impairment may worsen substantially over time [9]. In T2DM adult patients, the cognitive impairment can be divided into three different stages based on the severity [10, 11]: diabetes-associated cognitive decrements, mild cognitive impairment (MCI) and dementia. Diabetes-associated cognitive decrements refer to subtle changes in cognitive function. MCI affects one or more cognitive domains with largely preserved activities of daily life [12]. Dementia is the most severe stage that is defined as acquired objective cognitive impairment affecting multiple cognitive domains and daily activities [13, 14].

Cognitive impairment is a shared abnormality between DM and many neurodegenerative and neuropsychiatric disorders, such as Alzheimer's disease (AD) and schizophrenia [15]. In DM, cognitive impairment can be prevented or slowed down by consistent blood sugar management and a healthy lifestyle. However, there is no effective therapeutic agent to treat the condition at present [16]. The pathological mechanism is not entirely clear. Despite the widespread view that hyperglycemia and insulin resistance are predominant risk factors, convincing evidence demonstrated that oxidative stress plays an important role in contributing to the development of cognitive impairment in DM [17–19]. High-fat diet (HFD) is shown to directly increase oxidative damage and impair cognitive and memory capacity [20]. The dysfunctional metabolism of blood glucose damages the basilar membranes of capillaries leading to a narrowed cavity in the cerebra, thus decreasing blood supply to the brain and resulting in oxygen-free radical injury [21]. Oxidative stress is observed in cognitive impairment by increased levels of reactive oxygen species (ROS) and malondialdehyde (MDA), and decreased phase II antioxidant enzymes activities of glutathione peroxidase (GPx), chloramphenicol acetyltransferase (CAT) and superoxide dismutases (SOD) [22, 23]. The brain is highly susceptible to such oxidative damage partially due to its high oxygen demand, and the fact that high amounts of polyunsaturated fatty acids are easily targeted by free radicals [24]. Consequently, persistent hypoxia and oxygen-free radical injury can lead to neuronal cell injury and apoptosis [25]. Insulin is shown to penetrate the blood–brain barrier (BBB), and protects neurons against excessive free radicals, and the action is related to the nuclear factor erythroid 2-related factor 2 (Nrf2) regulated pathway [25]. Noticeably, insulin shows no protective effect on Nrf2-knockdown PC12 cells after H_2O_2-induced damage, while it significantly alleviated damage in $Nrf2^{+/+}$ cells [25]. Furthermore, insulin was shown to inhibit oxidative stress in neuronal cells as a potential antioxidant agent through the Nrf2-Keap1/antioxidant response element (ARE) signaling pathway [26]. Thus, Nrf2 may play an important role to protect cells against oxidative stress as an oxidative stress-responsive transcription factor. Since the induction of Nrf2 is shown to ameliorate cognitive impairment in neurological disorders [27], Nrf2 has been considered an emerging therapeutic target for the prevention and treatment of cognitive impairment in DM.

In this chapter, we aimed to investigate the mechanisms of Nrf2 as an antioxidant master regulator to attenuate the pathological development of cognitive impairment in DM. We also reviewed current evidence on Nrf2 inducers sourced from natural products as promising therapeutic agents to help manage the pathological conditions and symptoms of cognitive impairment in DM.

2. Nrf2 acts as an important therapeutic target for cognitive impairment in DM

2.1 Ameliorate oxidative stress

Oxidative stress is constantly involved in the development and progression of cognitive impairment in DM. The persistent hyperglycemia, excessive lipid, and high-level advanced glycation end products (AGEs) all contribute to the excessive production of ROS [28]. Thus, ROS stress is considered a link between neurodegenerative diseases and T2DM, which induced oxidized DNA, RNA, protein, and lipid products that are used as disease progression marks in patients with both AD and T2DM [1]. The overproduction of ROS also triggers a prolonged state of inflammation through the high generation of white adipose tissue that secretes proinflammatory mediators [29], and subsequent neuronal death that affects the function of organ systems [30, 31]. The chronic oxidative stress and the associated neuroinflammation, in turn, can exacerbate abnormal insulin secretion, insulin action, and immune responses [32].

A number of studies suggested that a high-fat diet induced-adiposity and insulin resistance increased cerebral oxidative stress and downregulated Nrf2 signaling [2, 20, 33–35]. This might be the main reason that triggers the declined cognitive performance in the aged brain [20]. In addition, the diabetes experimental models suggested that enhancing Sirt1/Nrf2 signaling pathway activity prevented oxidative stress-induced neuronal injury, and thus improved cognitive function [2]. Nrf2 is a core transcription factor of antioxidative stress which regulates the cellular redox status. It interacts with ARE and plays a wide range of cytoprotective roles in the prevention of oxidative stress, neurological damage and inflammatory responses [36]. Nrf2 modulates the expression of more than 200 downstream genes encoding Phase II response enzymes during the oxidative challenge, including heme Oxygenase (HO-1), glutathione S-transferase (GST), catalase (CAT), superoxide dismutase (SOD), and NADPH quinone oxidoreductase (NQO1) [37]. The mechanisms of action to suppress oxidative stress by the activation of Nrf2 is mediated through the restored production of many downstream antioxidant genes which directly reduce the intracellular ROS accumulation [38]. The upregulation of Nrf2 target genes in the CNS protects neurons against oxidative insults as shown by a broad array of published studies [24, 39–45].

2.2 Suppress NLR family pyrin domain containing 3 inflammasome-induced neuroinflammation

The nod-like receptor pyrin containing 3 inflammasomes (NLRP3) is a multimeric protein complex containing cytosolic NLRP3, the adaptor protein ASC and caspase-1.

Hyperglycaemia and other repetitive stress stimuli assemble and activate the NLRP3 inflammasome which then triggers the activation of caspase-1 by proteolytic cleavage [46]. The latter converts pro-IL-1β into bioactive IL-1β and consequently initiates inflammatory responses. The sustained inflammatory responses can cause neurotransmitter dysfunction which subsequently leads to a decreased production of brain-derived neurotrophic factor (BDNF), a protein responsible for the learning and memory process [47]. Thereby, it results in neuroinflammation-mediated neuronal damage as well as deterioration of cognitive function [48–50]. The aggravated neurodegeneration caused by the accumulation of IL-1β can be rescued by blocking IL-1β signaling in APP/PS1 mice [50–53], suggesting NLRP3 inflammasome-mediated neuroinflammation is an important therapeutic target. Moreover, inflammatory responses can also induce the overproduction of ROS which then exacerbate the oxidative damage to neurons [54].

Recent studies have shown that Nrf2/ARE signaling is directly implicated in the regulation of inflammation rather than relying on regulating oxidative stress to control inflammatory response [48, 49, 55, 56]. Although not fully understood, Nrf2 regulated AGEs/RAGE pathway is shown to mediate the activation of NLRP3 inflammation by providing a priming signal at a transcriptional level [57, 58] which is essential for the assembly of the NLRP3 inflammasome [59]. Excessive activation of this process exhausts cytosolic NADPH, increases the accumulation of ROS, and disrupts Nrf2/ARE pathway, finally leading to oxidative stress [60]. Besides, several studies revealed that suppressed intracellular ROS accumulation by the activation of Nrf2 resulted in reduced activation of the NLRP3 inflammasome, and subsequently inhibited neuroinflammation [38]. Additionally, Nrf2 activation triggers the restored expressions of HO-1 and NQO1 which then further contribute to the suppression of inflammation-associated cytokine productions and NLRP3 inflammasome formation [61–63].

2.3 Alleviate endoplasmic reticulum stress

The endoplasmic reticulum (ER) is involved in glucose-modulated secretory protein folding, intracellular Ca2+ storage, synthesis of unsaturated fatty acids, sterols, and phospholipids [64, 65]. High glucose levels in the body disrupt the internal balance of the ER, resulting in the accumulation of unfolded or misfolded proteins, which leads to ER stress (ERS) [66]. Oxidative stress also can induce ER stress responses [67] which are triggered by the accumulation of ROS [68].

Excessive ERS impairs neuronal function through various pathways. In hippocampal cells, high glucose and palmitic acid-induced ERS significantly decrease proBDNF, but were restored when ER stress was reduced [69]. In the neuroinflammatory pathway, ERS can activate the NLRP3 inflammasome and lead to the subsequent secretion of proinflammatory cytokines such as IL-1β [67, 68, 70] which is detrimental to the neuronal function as described above. In addition, ERS can further impair neuronal survival. The double-stranded RNA-dependent protein kinase (PKR)-like ER-resident kinase (PERK) pathway is one of the three ERS-related protective signaling pathways identified so far. PERK pathway can activate caspase-12, which is a specific mediator of ER stress-induced apoptosis, to impair protein synthesis and synaptic functions [70]. Excessive or prolonged ER stress could up-regulate the transcription factor C/EBP Homologous Protein (CHOP), which is a critical mediator of the Bax/Bcl-2 -dependent pathway, thus converting neuronal cells from the pro-survival to pro-apoptosis phase [71, 72]. Thus, ERS is considered a valuable therapeutic target in neuroprotection such as the cognitive decline in DM [73].

ERS can cause and exacerbate oxidative stress, whereas oxidative stress can worsen ERS. In response to ERS, the onset of the PERK pathway quickly activates Nrf2 as an antioxidant defensive system [66], although the clear link remains to be explored. As an endogenous defense system against oxidative stress, Nrf2 regulates the expression of a subset of detoxifying enzymes including NQO1 and γ-GCS. This suggests an important cytoprotective role of Nrf2 against ERS, and that the Nrf2 pathway represents a useful therapeutic strategy in cognitive decline by attenuating ERS. However, the moderate self-defense response of PERK and Nrf2 maybe not be enough to prevent damage [73]. The other pathway that Nrf2 may exert a beneficial effect in attenuating ERS is that the Nrf2 activation contributes to the maintenance of glutathione levels, which, in turn, functions as a buffer for the accumulation of ROS during the unfolded protein response in ER and thus may attenuate ERS. In addition, the reduced level of ROS may reduce apoptotic induction as a consequent event after ERS [74].

2.4 Repair a leaking BBB

Impairment of BBB induced by hyperglycemia, oxidative damage, and inflammation is a critical neurovascular complication of DM that adversely affects the microvascular environment, health, and function of the central nervous system. Many studies have demonstrated that hyperglycemia-driven neuroinflammation is a risk factor for BBB disruption leading to a high permeability of BBB [75–77]. The BBB leakage leads to many adverse impacts on the central nervous system including decreased waste transport, increased infiltration of immune cells and subsequent glial and neuron dysfunction, over-active immune sensitivity, hormone dysregulation, and cognitive impairment [78]. The excessive production of ROS also induces endothelial oxidative stress and mitochondrial damage which has been recognized as a central pathological mechanism for BBB dysfunction in DM.

A study from Sajja et al. suggested a close link of Nrf2 protein expression to that of the BBB permeability, as evidenced by a significantly down-regulated brain Nrf2 protein expression in the db/db diabetic mice with a strong increase in BBB permeability (to 70 kDa dextran) [79]. The role of Nrf2 induction in the repairment of BBB appeared to be associated with the mitochondrial transporter *ABCB10* which is considered as an essential player in mitochondrial function and redox balance at BBB endothelium. The *ABCB10* knockdown resulted in a strong induction of Nrf2-driven antioxidant responses manifested as increased expression of Nrf2 and its downstream antioxidant genes [79]. In addition, several studies suggested that Nrf2-driven free radical detoxification pathways are essential in protecting against oxidative stress-induced endothelial injuries [80], and thus may in turn contribute to the protection of BBB [81].

2.5 Protect mitochondria function

Mitochondrial dysfunction is implicated in the pathogenesis of most neurodegenerative disorders including cognitive impairment in DM [82]. Mitochondria is the main source of energy in the cell, by providing ATP through oxidative phosphorylation and harboring several metabolic pathways such as fatty acid oxidation (FAO) and the tricarboxylic acid cycle (TCA) cycle [82]. Accumulated oxidative damage causes an altered oxidative phosphorylation and redox imbalance in mitochondria [83]. The long-lasting effect can lead to the des-regulation of mitochondrial protein homeostasis, and consequent proteotoxic stress in the neurons [84]. Neurons in the central nervous system have a high requirement for energy as attributed to their constant activity and signaling, and thus are especially vulnerable to mitochondrial dysfunction. Decreased ATP production is a common hallmark of neurodegenerative diseases and can be caused by various mechanisms, such as the impaired activity of any of the complexes of the respiratory chain, alterations in glucose uptake, glycolysis, TCA cycle, or uncoupling [85].

Nrf2 can lead to mitochondrial bioenergetic improvement, although the mechanisms of action are not fully explored [82]. There are several mechanisms of action that underpin the role of Nrf2 in protecting mitochondrial function. Nrf2 prevents the oxidative thiol modifications that can modulate the function of proteins implicated in metabolic pathways [86–88]. The accumulative mitochondrial ROS can reversibly modify thiol groups presented in several enzymes implicated in carbohydrate and lipid metabolism. The activation of Nrf2 that scavenge or inhibit the overproduction of mitochondrial ROS may prevent the redox modifications of the sensitive thiol groups [89–91]. Consequently, such antioxidant modification enables the restoration of normal glycolytic metabolism and reduces neuronal apoptosis [92].

2.6 Modulation of glucose metabolism

Diabetic patients (both type 1 and type 2) have a higher risk of AD and vascular dementia, mainly caused by abnormal glucose metabolism induced by hyperglycemia and consequent cerebrovascular lesions in the frontal lobe [93]. The elevated glucose concentration leads to enhanced oxidative phosphorylation and increased levels of glutamate, an excitatory neurotransmitter, whose enhanced levels provoke cognitive dysfunction by causing neuronal damage.

Although the majority of evidence stands for the antioxidant properties of Nrf2 activation, Nrf2 is also involved in glucose metabolism [94, 95]. Nrf2 can directly activate the transcription of enzymes containing ARE-sequences in the gene promoters, which have been described in the promoters of the genes encoding the peroxisome proliferator-activated receptor γ (PPAR- γ) [96]. PPAR-γ is known to regulate adipocyte differentiation and lipid metabolism, as well as the network regulation of glucose homeostasis [97]. It also exhibited a versatile role in early brain development and post-injury brain repair [98]. The activation of PPAR- γ is shown to repair damaged tissue through angiogenesis, alleviate neurological deficits, and exhibit neuroprotective activity [96]. Indeed, studies have revealed the connection of Nrf2 and PPAR-γ *via* a positive feedback loop that simultaneously strengthens each other's expression [94, 95]. For example, PPAR-γ expression was found to be significantly lower in Nrf2 knockout mice [99]. It is also identified that one of the many Nrf2 downstream targets, *TALDO1*, contains an ARE sequence which is a key enzyme in the nonoxidative pentose phosphate pathway, providing ribose-5-phosphatase and NADPH for nucleic acid and lipid biosynthesis [92].

Taken together, the recently recognized abilities of Nrf2 to ameliorate oxidative and ERS, repair BBB, protect mitochondrial function and regulate glucose metabolism make the Nrf2 activation an attractive and comprehensive therapeutic target for

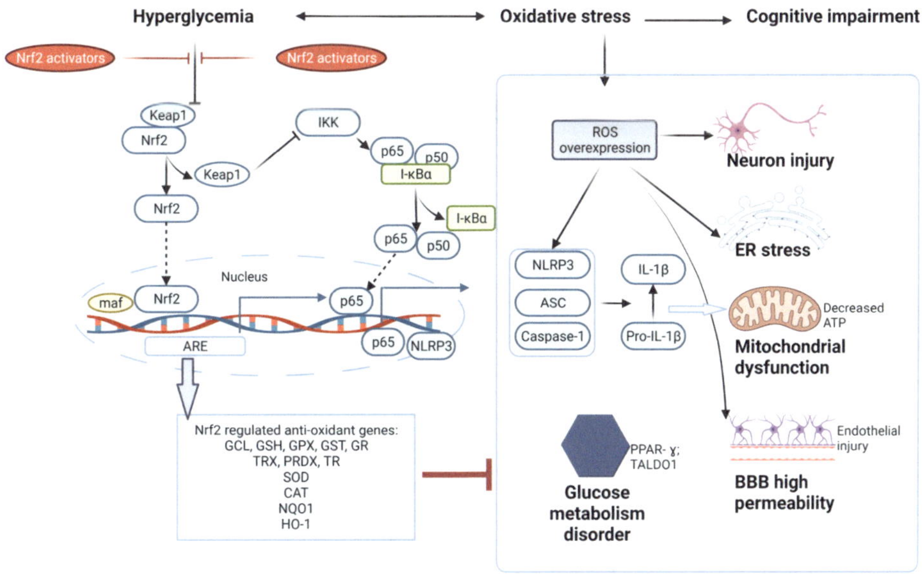

Figure 1.
A summarized diagram of Nrf2-associated pathways in the pathological mechanisms of DCI.

the management of cognitive impairment in DM. A summarized diagram of Nrf2-associated pathways in the pathological mechanisms is shown in **Figure 1**.

3. Nrf2 activators from natural products that prevent or treat cognitive impairment in DM

Natural products have served as a rich source of novel drug discovery and development. This section focuses on the promising Nrf2 inducers sourced from natural products that have shown beneficial effects in the prevention or treatment of cognitive impairment in DM. A literature search was conducted in scientific databases (PubMed, MEDLINE and EMBASE) and Google Scholar for English-published studies between January 2000 and September 2022. Clinical and pre-clinical studies of natural products (i.e., herbal extracts, vegetables, plants, chemicals, nutraceuticals, and supplements sourced from plant or plant extract) that showed effects in attenuating cognitive impairment in DM and the mechanism of action was associated with the Nrf2 activation were included. Studies investigating pharmaceutical drugs, analog drugs, and synthesized compounds were excluded.

Nineteen preclinical studies demonstrated that various natural compounds attenuated cognitive impairment in DM at least partially through the induction of Nrf2 (see **Table 1**). The effects on cognitive impairment by the natural products were mostly demonstrated in the diabetic animal models, either on high-fat diet (HDF) or streptozotocin (STZ) injection-induced diabetic rats [111, 113, 114], db/db mice (model phase 1 to 3 of diabetes type II and obesity) [100, 115], or diabetic Goto-Kakizaki rats (non-obese type 2 diabetic model) [110]. One cellular study was included which utilized chronic high glucose-exposed SHSY5Y neurons [116]. The studies on natural Nrf2 activators with strong preclinical evidence and associated molecular mechanisms are discussed in detail below.

Natural compounds	Source	Administration	Key results	Mechanism related to Nrf2 activation
Sulforaphane [100, 101]	Vegetables	1 mg/kg, i.p. for 28 days	Mitigated cognitive impairment of db/db mice; decreased Aβ oligomers and plaques, phosphor-tau and Thr231 in hippocampus	↑Nrf2 → ↑HO-1 and NQO1 → ↓ROS level
		25 mg/kg, orally once daily for consecutive 14 days	Prevented the memory impairment, decreased the apoptosis of hippocampal neurons in STZ-injected SD rats	↓ Caspase-3 and myeloid cell leukemia 1 (MCL-1); ↑ p-Akt, p-GSK3β, NGF and BDNF
Troxerutin [102, 103]	*Sophora japonica*	60 mg/kg, 1 mL/kg, i.p., 12 weeks	Improved learning and memory levels in STZ-induced diabetic rats	↑Nrf2 → ↑SOD and ↓MDA
		150 mg/kg/day intraperitoneally for 6 weeks	Improved cognitive impairment in STZ-induced diabetic rats	↑ Nrf2 translocation → ↑ HO-1 and NQO1 → NOX subunits → ↓MDA and ↑SOD
Strawberry leaf extract [104]	Strawberry tree	200 mg/kg for 4 weeks	Alleviated cognitive impairment in STZ-induced diabetic rats	↑Nrf2 -HO-1 signaling → ↓ROS, ↓MDA, ↑SOD and CAT; ↓IL-6 and TNF-α; ↓caspase-3 and caspase-9 in hippocampus
Betulin [105]	Birch tree bark	20 or 40 mg/kg for 4 weeks	Improved glucose intolerance and modify basal learning performance in STZ-induced diabetic rats	↑Nrf2 -HO-1 signaling and ↓NF-κB → ↑SOD and ↓MDA; ↓inflammatory cytokines in serum and hippocampus
S-allyl cysteine [106]	Aged garlic extract	150 mg/kg/day for 7 weeks (p.o.)	Lowered serum glucose, improved spatial recognition memory, discrimination ratio in novel object recognition task, and restored step-through latency (STL) in passive avoidance paradigm in STZ-induced diabetic rats	↑Nrf2-HO-1 signaling and ↓TLR4-NF-κB signaling → ↓ acetylcholinesterase activity, MDA; ↑SOD and GSH
Caffeic acid phenethyl ester [107]	Propolis	200 and 400 mg/kg p.o. for 14 days	Rescued the diabetic brain atrophy and diminish CA1 and CA3 cells of hippocampus and cerebral cortex in STZ-induced diabetic mice; decrease Aβ and p-tau (S396)	↑Nrf2-related anti-oxidation mechanisms → antioxidation, anti-inflammation and autophagy induction
AB-38b [108]	Fructus *Schisandrae chinensis*	0, 10, 20 and 40 mg/kg by gavage for 8 weeks	Increased the preference index to novel object and the number of neurons in hippocampal CA1 area of diabetic mice	↑Nrf2 expression and phosphorylation → γ-GCS

Natural compounds	Source	Administration	Key results	Mechanism related to Nrf2 activation
Fish oil supplementation [109]	Fish oil	1.5 g/kg/d (34% EPA, 24% DHA) for 10 weeks	Improved spatial learning and memory in STZ-induced diabetic rats	Nrf2 and HO-1 in cortex and hippocampus → ↓oxidative stress → ↓ IL-1β, IL-6, and TNF-α, ↑ IL-4 and IL-10
Soy isoflavones [110]	Soy	20 mg/kg once a day for 4 weeks	Alleviated the cognitive dysfunction of the diabetic Goto-Kakizaki rats	↑ Nrf2, HO-1 and NQO1 → oxidative reactions
Resveratrol [111, 112]	Grapes, berries, etc.	30 mg/kg every other day for 4 months by intragastric administration	Prevented the learning and memory decline and hippocampal neuron destruction and synaptic ultrastructural damage in HFD and STZ- induced T2DM mice	↑ Nrf2 → ↑ HO-1 and NQO1 proteins; ↑ SOD and CAT → ↓ TNF-α and IL-1β
		20 mg/kg, intraperitoneally once daily for 4 weeks	Significantly elevated total oxidant species (1.22-fold) and Malonedialdehyde (MDA) (1.38-fold) contents in diabetic rat brain cortex tissues in STZ (55 mg/kg)-injected Wistar rats	↓oxidative conditions
Thymol (2-isopropyl-5-methylphenol) [113]	Thyme	20, 40 mg/kg for 12 weeks	Reversed the gain of body weight and peripheral insulin resistance induced by HFD; improved the cognitive impairments; decreased HFD-induced Aβ deposition and tau hyperphosphorylation	↑ Nrf2-HO-1 signaling → ↓ oxidative stress and inflammation
Astaxanthin [114]	Algae, yeast, salmon, trout, krill, shrimp, and crayfish	25 mg/kg 3 times/week for 6 weeks by intraperitoneally	Ameliorated the impairment in the neurons of HFD and STZ-induced diabetic rats	↑Nrf2-HO-1 and ↓NF-κB signalings → ↑ SOD and ↓ MDA; ↓IL-1β and IL-6
Notoginsenoside R1 [115]	*Panax notoginseng*	10, 30 mg/kg once daily administrated intragastrically for 10 weeks	Ameliorated cognitive dysfunction, depression-like behaviors, insulin resistance, hyperinsulinemia, dyslipidemia, and inflammation in db/db mice	↑Akt → ↑ Nrf2 and ↓ NLRP3
Albiflorin [7]	*Paeonia lactiflora*	100, 200 mg/kg by gastric gavage for 10 weeks	Improved spatial and learning ability; decrease in Aβ plaque density in the hippocampus of rats;	Regulate Nrf2/HO-1/HMGB1/NF-κB → ↓ oxidative stress and inflammation

Natural compounds	Source	Administration	Key results	Mechanism related to Nrf2 activation
Quercetin [116, 117]	Fruits and vegetables	5, 10, 20 μmol/L for 72 h	Increased cell viability, and enhanced Glo-1 functions in SHSY5Y neurons	↑Nrf2 and p-Nrf2 levels → ↑ γ-GCS protein and mRNA
		50 mg/kg i.p. for 18 days	Significantly mitigated the STZ-induced increase in cholinergic dysfunction in STZ (3 mg/kg)-induced brain mitochondrial toxicity in Alzheimer's disease -like rats	↑α7nAChR/Nrf2/HO-1-mediated neuroprotection

Table 1.
Summarized studies for natural products that attenuate cognitive impairment in DM and associated symptoms via the mechanism of Nrf2 activation.

3.1 Sulforaphane

Sulforaphane is a hydrolysis product of glucoraphanin, a group of sulfur-containing glycosides, found in many raw vegetables such as broccoli, cauliflower, bok choy, and watercress [118–121]. Upon consumption, glucoraphanin is hydrolyzed to sulforaphane which has been demonstrated to penetrate BBB [122] and exhibit a neuroprotective effect *via* a number of mechanisms [123]. In particular, strong evidence has indicated that sulforaphane is a potent Nrf2 activator. The molecular mechanisms are linked with increased Nrf2 transcription by reducing methylation of the first 15 CpGs of Nrf2 promoters [124], modulation of Kelch-like ECH-associated protein 1 (Keap1) [125–127], prevention of ubiquitination of Nrf2 [128], and enhanced Nrf2 translocation [128]. The activation of Nrf2 by sulforaphane was reported to mediate a long-lasting effect of the upregulation of cytoprotective enzymes (4-hour exposure of sulforaphane led to 24 hrs NQO1 and HO-1 mRNA elevation and over 48 hrs increase of corresponding proteins) [129].

Two *in vivo* studies have investigated the effect of IP/oral injection of sulforaphane on mitigating cognitive impairment in diabetic models [100, 101]. In particular, the study by Pu et al. suggested that the IP injection of sulforaphane (1 mg/kg for 28 days) improved the spatial learning and memory function of db/db mice examined by the Morris water maze tests. Sulforaphane was shown to decrease the levels of Aβ oligomers, Aβ 1–42 plaques, phosphor-tau at Ser396 and Thr231 in the hippocampus. The molecular mechanisms were related to Nrf2-mediated upregulation of HO-1 and NQO1 protein expression, and reduced ROS/RNS levels [100]. In line with the research finding, a study by Wang et al. revealed that the oral administration of sulforaphane (25 mg/kg) for consecutive 14 days prevented memory impairment in STZ-induced diabetic rats [101]. It was revealed that sulforaphane markedly reduced apoptotic neurons levels and it was associated with decreased phospho-protein kinase B (Akt), phosphor-glycogen synthase kinase 3β (GSK3β), nerve growth factor (NGF) and BDNF expressions. However, the link between the reduced neuronal apoptotic level and the Nrf2 activation remains to be explored. Broadly speaking, many studies have agreed the strong neuroprotective effect of sulforaphane by activating Nrf2 antioxidant signaling cascade, and such activity can be used in many neurodegenerative diseases including AD, traumatic brain injury, and ischemic stroke [130]. Future studies are suggested to explore the safety profile and pharmacokinetic characteristics of sulforaphane to further demonstrate its therapeutic value in improving cognitive function in humans.

3.2 Astaxanthin

Astaxanthin is a red pigment belonging to a group of carotenoids that causes the pink-red color in salmon. Astaxanthin has various biological and pharmacological activities such as antioxidant, anti-inflammatory, and antidiabetes [131]. Many studies demonstrated that astaxanthin exhibited potential neuroprotective effects, of which the mechanism was associated with Nrf2 activation [131–134]. Astaxanthin administration was shown to accelerate nuclear translocation of Nrf2 *via* inducing PI3K/Akt signaling, resulting in a decrease in ROS levels, and inhibiting apoptosis in neuronal cells [135].

Orally or parenterally administered astaxanthin has been shown to improve insulin resistance and insulin secretion, attenuate hyperglycemia, and exert protective effects against many DM complications including retinopathy, nephropathy, and neuropathy [136]. Specifically, for cognitive impairment in DM, astaxanthin administration improves cognitive function [137] and diminish oxidative stress, nitric oxide synthase, and inflammation in DM rat [137]. Furthermore, Feng et al. suggested that IP injection of astaxanthin (25 mg/kg) ameliorated the impairment in the neurons of HFD and STZ-induced diabetic rats, and the action was associated with upregulated the Nrf2-HO-1 signaling pathway which drove the increased expressions of SOD and declined MDA. The intermediate metabolites from the Nrf2-HO-1 axis contribute to the suppressed inflammatory response by inhibiting the activation of NF-κB leading to decreased levels of IL-1β and IL-6 [114].

3.3 Resveratrol

Resveratrol (3,4′,5-trihydroxy-stilbene, RES) is a very popular nutraceutical that is a naturally occurring compound found in grapes, cereals, vegetables, dry legumes and plant-derived beverages, including tea, coffee, and wine. A large number of studies supported its beneficial use in anti-diabetes, anti-inflammatory, and antioxidative conditions [138–142]. In diabetes, resveratrol is shown to exhibit an insulin-sensitizing effect, enhance glucose uptake and metabolism, preserve islet β-cells and release insulin from β-cells [138]. Among all the associated molecular mechanisms of resveratrol in treating diabetic complications, one of the key pathways was the upregulation of Nrf2 nuclear translocation which then induced the increased expressions of GSH, SOD, NQD1, and HO-1, reduced ROS production, and declined oxidative stress [138]. Two studies specifically investigated the protective effect of resveratrol on cognitive function in diabetic rats *via* regulating oxidative biomarkers and antioxidant enzymes in the brain [111, 112]. A study from Sadi et al. suggested that resveratrol significantly elevated total oxidant species in diabetic rat brain cortex tissues in STZ-injected Wistar rats [112]. Moreover, Wang et al. showed that resveratrol prevented the learning and memory decline in STZ-induced T2DM mice [111], and such action was linked with upregulated Nrf2 and its mediated HO-1 and NQO1 proteins, SOD and CAT enzymes, as well as reduced IL-1β and TNF-α. A randomized, double-blind, placebo-controlled clinical trial demonstrated that a total of 48 patients with T2D received 800 mg/day of resveratrol for 2 months markedly increased Nrf2 and SOD expressions in peripheral blood mononuclear cells. Resveratrol was well tolerated without causing any serious adverse events [143]. Thus, it is encouraging that resveratrol may exhibit beneficial clinical outcomes in T2D patients with cognitive decline as a stronger Nrf2 activator.

3.4 Quercetin

Quercetin is one of the most abundant polyphenolic flavonoids, which is present in fruits and vegetables and displays a strong health-promoting effect [144]. In addition, quercetin plays an indirect role in neutralizing oxidative stress by activating the Nrf2-ARE antioxidant pathway and inducing the expression of antioxidant enzymes like CAT and SOD. In lipopolysaccharides (LPS)-induced murine BV-2 microglial cells, quercetin produced a greater stimulating effect on Nrf2-induced increase expression of heme-oxygenase-1 (HO-1) protein than cyanidin against endotoxic stress *via* the participation of mitogen-activated protein kinase (MAPKs) [145, 146]. In particular, quercetin was found to increase cell viability and enhance glyoxalase-I (Glo-1) functions in SHSY5Y cells which were related to activated Nrf2 and phosphor-Nrf2 levels [116]. Furthermore, IP injection of quercetin (50 mg/kg) for 18 days significantly mitigated the STZ-induced increase in cholinergic dysfunction in STZ (3 mg/kg)-induced brain mitochondrial toxicity in AD-like rats, which the mechanism was associated with α7nAChR/Nrf2/HO-1-mediated neuroprotection [117]. However, a clinical trial is warranted to further investigate the effect of quercetin in humans.

4. Conclusion and future perspectives

A number of natural compounds, such as sulforaphane, astaxanthin, resveratrol, and quercetin [115] were found to significantly improve the spatial and memory function in diabetic animals models (rats/mice) with the effective daily dosage ranging from 1 mg/kg to 30 mg/kg. Most of the treatments were given after the induction of DM. In addition, most treatments from natural products were long-term treatment duration (14 days to 4 months) with a daily administration to see an obvious effect. The summarized information is listed in **Table 1**. However, clinical trials are largely lacking to further demonstrate their safety and efficacy in humans.

Nrf2 inducers sourced from natural products upregulated Nrf2 activity, Nrf2 translocation (nuclear Nrf2 protein expression) and/or phosphorated Nrf2 expression. This leads to the activation of Nrf2-regulated downstream antioxidant genes including HO-1, NQO1, SOD and CAT [102–104, 106, 111, 114]. The reduced levels of ROS and MDA in the hippocampus were often reported as the consequent events. Interestingly, many studies also mentioned the inhibited NF-κB signaling in addition to the Nrf2-mediated antioxidant response [7, 106, 114]. The upregulated expression of HO-1 protein in response to the Nrf2 upregulation suppresses the NF-κB activation which governs the transcriptions and productions of proinflammatory cytokines. Thus, Nrf2/HO-1 pathway not only drives the antioxidant response but also plays a critical role in regulating the proinflammatory reactions. The suppressed proinflammatory response in the animals is evidenced by the reduced production of IL-6 and TNF-α and increased levels of IL-4 and IL-10 [109, 114]. However, the inner relationship among Nrf2-mediated antioxidants and anti-inflammatory by the Nrf2 activators is not fully explored.

It is worth mentioning that the Nrf2 activation in cognitive impairment was shown to be mediated by the induction of protein kinase B (Akt) as an upstream mediator.

A study from Zhai et al. suggested that notoginsenoside R1 (NR1) activated the Nrf2 pathway as well as inhibited NLRP3 inflammasome which both contributed to ameliorated cognitive function. However, the neuroprotective effect after the Nrf2 activation and NLRP3 inflammasome inhibition was abolished by the co-treatment of Akt inhibitor and NR1 [115]. It has been well-studied that the activation of Akt regulates cell survival and apoptosis as well as glucose transport [147, 148]. NR1 was capable to improve glucose tolerance and insulin sensitivity in db/db mice and reducing neuron apoptosis in the hippocampal CA1 region, and the mechanisms may be associated with the activation of the Akt and Nrf2 signaling cascade [115]. These two studies highlight the possible molecular action of Nrf2 inducers mediated by the Akt pathway in protecting neuronal survival in the hippocampal region. Further studies are warranted to elucidate the link between Akt and Nrf2 pathways.

Taken together, this book chapter introduces cognitive impairment in DM as the major complication in both type 1 and type 2 DM, and explained the associated pathological mechanisms, particularly the detrimental role of oxidative stress that impairs cognitive and memory capacity. Nrf2, the core intracellular transcription factor of antioxidative stress, has been considered an emerging therapeutic target against cognitive impairment in DM. The induction of Nrf2 signaling drives the antioxidant defense system to ameliorate oxidative and ER stress, repair the BBB, protect mitochondrial function, and regulate glucose metabolism, which eventually protects neurons against oxidative stresses and hyperglycemia and helps to restore cognitive function. A number of studies, predominantly based on diabetic animal models, have demonstrated that Nrf2 inducers sourced from natural products such as vegetables, soy, garlic, and strawberry leaf improved cognitive and memory function. The associated mechanisms of action were shown as Nrf2-mediated antioxidant and anti-inflammatory responses. Although the results are supportive, conclusive evidence for each proposed therapeutic candidate is weak due to the low quantity of studies. The safety human trials to further demonstrate their safety and efficacy are warranted. However, the findings of this chapter may shed light on the development of natural Nrf2 activators as promising pharmacological agents to prevent or slow the progress of cognitive impairment in DM.

Acknowledgements

I am grateful to Professor Gerald Münch who inspires me greatly in the research field of the application of Nrf2-mediated antioxidant defense system in the prevention and treatment of many diseases including neurodegenerative diseases. Thanks to Mr. John Truong for his kind support.

Conflict of interest

The authors declare no conflict of interest. As a medical research institute, NICM Health Research Institute receives research grants and donations from foundations, universities, government agencies, individuals and industry. Sponsors and donors also provide untied funding for work to advance the vision and mission of the Institute.

Author details

Xian Zhou[1*], Gerald Münch[2] and Dennis Chang[1]

1 NICM Health Research Institute, Western Sydney University, Westmead, MSW, Australia

2 Pharmacology Unit, Western Sydney University, Campbelltown, NSW, Australia

*Address all correspondence to: p.zhou@westernsydney.edu.au

References

[1] He L, Sun Y. The potential role of Keap1-Nrf2 pathway in the pathogenesis of Alzheimer's disease, type 2 diabetes, and type 2 diabetes-related Alzheimer's disease. Metabolic Brain Disease. 2021;**36**:1469-1479. DOI: 10.1007/s11011-021-00762-z

[2] Han J, Liu X, Li Y, Zhang J, Yu H. Sirt1/Nrf2 signalling pathway prevents cognitive impairment in diabetic rats through anti-oxidative stress induced by miRNA-23b-3p expression. Molecular Medicine Reports. 2018;**17**:8414-8422. DOI: 10.3892/mmr.2018.8876

[3] Vinik AI, Nevoret M-L, Casellini C, Parson H. Diabetic neuropathy. Endocrinology and Metabolism Clinics of North America. 2013;**42**:747-787. DOI: 10.1016/j.ecl.2013.06.001

[4] Figueroa-Romero C, Sadidi M, Feldman EL. Mechanisms of disease: The oxidative stress theory of diabetic neuropathy. Reviews in Endocrine & Metabolic Disorders. 2008;**9**:301-314. DOI: 10.1007/s11154-008-9104-2

[5] Degen C, Toro P, Schönknecht P, Sattler C, Schröder J. Diabetes mellitus type II and cognitive capacity in healthy aging, mild cognitive impairment and Alzheimer's disease. Psychiatry Research. 2016;**240**:42-46. DOI: 10.1016/j.psychres.2016.04.009

[6] Luchsinger JA, Reitz C, Patel B, Tang M-X, Manly JJ, Mayeux R. Relation of diabetes to mild cognitive impairment. Archives of Neurology. 2007;**64**:570-575. DOI: 10.1001/archneur.64.4.570

[7] Ma X, Song M, Yan Y, Ren G, Hou J, Qin G, et al. Albiflorin alleviates cognitive dysfunction in STZ-induced rats. Aging (Albany NY). 2021;**13**:18287. DOI: 10.18632/aging.203274

[8] Ryan CM, van Duinkerken E, Rosano C. Neurocognitive consequences of diabetes. The American Psychologist. 2016;**71**:563. DOI: 10.1037/a0040455

[9] Nunley KA, Rosano C, Ryan CM, Jennings JR, Aizenstein HJ, Zgibor JC, et al. Clinically relevant cognitive impairment in middle-aged adults with childhood-onset type 1 diabetes. Diabetes Care. 2015;**38**:1768-1776. DOI: 10.2337/dc15-0041

[10] Koekkoek PS, Kappelle LJ, van den Berg E, Rutten GE, Biessels GJ. Cognitive function in patients with diabetes mellitus: Guidance for daily care. Lancet Neurology. 2015;**14**:329-340. DOI: 10.1016/S1474-4422(14)70249-2

[11] Dolores EB. Diagnostic and statistical manual of mental disorders. Codas. 2013;**21**:591-643. DOI: 10.1590/s2317-17822013000200017

[12] Petersen RC. Mild cognitive impairment as a diagnostic entity. Journal of Internal Medicine. 2004;**256**:183-194. DOI: 10.1111/j.1365-2796.2004.01388.x

[13] Petersen RC, Lopez O, Armstrong MJ, Getchius TS, Ganguli M, Gloss D, et al. Practice guideline update summary: Mild cognitive impairment: Report of the guideline development, dissemination, and implementation Subcommittee of the American Academy of neurology. Neurology. 2018;**90**:126-135. DOI: 10.1212/WNL.0000000000004826

[14] Dubois B, Feldman HH, Jacova C, Cummings JL, DeKosky ST, Barberger-Gateau P, et al. Revising

the definition of Alzheimer's disease: A new lexicon. Lancet Neurology. 2010;**9**:1118-1127. DOI: 10.1016/S1474-4422(10)70223-4

[15] Chen Q, Cao T, Li N, Zeng C, Zhang S, Wu X, et al. Repurposing of anti-diabetic agents as a new opportunity to alleviate cognitive impairment in neurodegenerative and neuropsychiatric disorders. Frontiers in Pharmacology. 2021;**12**:667874. DOI: 10.3389/fphar.2021.667874

[16] Sharma G, Parihar A, Talaiya T, Dubey K, Porwal B, Parihar MS. Cognitive impairments in type 2 diabetes, risk factors and preventive strategies. Journal of Basic and Clinical Physiology and Pharmacology. 2020;**31**:20190105. DOI: 10.1515/jbcpp-2019-0105

[17] Muriach M, Flores-Bellver M, Romero FJ, Barcia JM. Diabetes and the brain: Oxidative stress, inflammation, and autophagy. Oxidative Medicine and Cellular Longevity. 2014;**2014**:102158. DOI: 10.1155/2014/102158

[18] Domínguez R, Pagano M, Marschoff E, González S, Repetto M, Serra J. Alzheimer disease and cognitive impairment associated with diabetes mellitus type 2: Associations and a hypothesis. Neurología. 2014;**29**:567-572. DOI: 10.1016/j.nrl.2013.05.006

[19] Hoyos CM, Colagiuri S, Turner A, Ireland C, Naismith SL, Duffy SL. Brain oxidative stress and cognitive function in older adults with diabetes and pre-diabetes who are at risk for dementia. Diabetes Research and Clinical Practice. 2022;**184**:109178. DOI: 10.1016/j.diabres.2021.109178

[20] Morrison CD, Pistell PJ, Ingram DK, Johnson WD, Liu Y, Fernandez-Kim SO, et al. High fat diet increases hippocampal oxidative stress and cognitive impairment in aged mice: Implications for decreased Nrf2 signaling. Journal of Neurochemistry. 2010;**114**:1581-1589. DOI: 10.1111/j.1471-4159.2010.06865.x

[21] Aruoma OI, Neergheen VS, Bahorun T, Jen L-S. Free radicals, antioxidants and diabetes: Embryopathy, retinopathy, neuropathy, nephropathy and cardiovascular complications. Neuroembryology Aging. 2006;**4**:117-137. DOI: 10.1159/000109344

[22] Padurariu M, Ciobica A, Hritcu L, Stoica B, Bild W, Stefanescu C. Changes of some oxidative stress markers in the serum of patients with mild cognitive impairment and Alzheimer's disease. Neuroscience Letters. 2010;**469**:6-10. DOI: 10.1016/j.neulet.2009.11.033

[23] Tayanloo-Beik A, Kiasalari Z, Roghani M. Paeonol ameliorates cognitive deficits in streptozotocin murine model of sporadic Alzheimer's disease via attenuation of oxidative stress, inflammation, and mitochondrial dysfunction. Journal of Molecular Neuroscience. 2022;**72**:336-348. DOI: 10.1007/s12031-021-01936-1

[24] Vasconcelos AR, Dos Santos NB, Scavone C, Munhoz CD. Nrf2/ARE pathway modulation by dietary energy regulation in neurological disorders. Frontiers in Pharmacology. 2019;**10**:33. DOI: 10.3389/fphar.2019.00033

[25] Song Y, Ding W, Bei Y, Xiao Y, Tong H-D, Wang L-B, et al. Insulin is a potential antioxidant for diabetes-associated cognitive decline via regulating Nrf2 dependent antioxidant enzymes. Biomedicine & Pharmacotherapy. 2018;**104**:474-484. DOI: 10.1016/j.biopha.2018.04.097

[26] Bhakkiyalakshmi E, Sireesh D, Rajaguru P, Paulmurugan R, Ramkumar KM. The emerging role of

redox-sensitive Nrf2–Keap1 pathway in diabetes. Pharmacological Research. 2015;**91**:104-114. DOI: 10.1016/j.phrs.2014.10.004

[27] Uruno A, Matsumaru D, Ryoke R, Saito R, Kadoguchi S, Saigusa D, et al. Nrf2 suppresses oxidative stress and inflammation in App knock-in Alzheimer's disease model mice. Molecular and Cellular Biology. 2020;**40**:e00467-e00419. DOI: 10.1128/MCB.00467-19

[28] Ahmad W, Ijaz B, Shabbiri K, Ahmed F, Rehman S. Oxidative toxicity in diabetes and Alzheimer's disease: Mechanisms behind ROS/RNS generation. Journal of Biomedical Science. 2017;**24**:1-10. DOI: 10.1186/s12929-017-0379-z

[29] Knight JA. Free radicals, antioxidants, and the immune system. Annals of Clinical and Laboratory Science. 2000;**30**:145-158

[30] Zhao Y, Zhao B. Oxidative stress and the pathogenesis of Alzheimer's disease. Longevity OMAC. 2013;**2013**:316523. DOI: 10.1155/2013/316523

[31] Selfridge JE, Lezi E, Lu J, Swerdlow RH. Role of mitochondrial homeostasis and dynamics in Alzheimer's disease. Neurobiology of Disease. 2013;**51**:3-12. DOI: 10.1016/j.nbd.2011.12.057

[32] Newsholme P, Cruzat VF, Keane KN, Carlessi R, de Bittencourt JPIH. Molecular mechanisms of ROS production and oxidative stress in diabetes. The Biochemical Journal. 2016;**473**:4527-4550. DOI: 10.1042/BCJ20160503C

[33] Zhang Z, Yang H, Yang J, Xie J, Xu J, Liu C, et al. Pseudoginsenoside-F11 attenuates cognitive impairment by ameliorating oxidative stress and neuroinflammation in d-galactose-treated mice. International Immunopharmacology. 2019;**67**:78-86. DOI: 10.1016/j.intimp.2018.11.026

[34] Xiang T, Luo X, Ye L, Huang H, Wu Y. Klotho alleviates NLRP3 inflammasome-mediated neuroinflammation in a temporal lobe epilepsy rat model by activating the Nrf2 signaling pathway. Epilepsy & Behavior. 2022;**128**:108509. DOI: 10.1016/j.yebeh.2021.108509

[35] Madhu LN, Kodali M, Attaluri S, Shuai B, Melissari L, Rao X, et al. Melatonin improves brain function in a model of chronic gulf war illness with modulation of oxidative stress, NLRP3 inflammasomes, and BDNF-ERK-CREB pathway in the hippocampus. Redox Biology. 2021;**43**:101973. DOI: 10.1016/j.redox.2021.101973

[36] Singh S, Vrishni S, Singh BK, Rahman I, Kakkar P. Nrf2-ARE stress response mechanism: A control point in oxidative stress-mediated dysfunctions and chronic inflammatory diseases. Free Radical Research. 2010;**44**:1267-1288. DOI: 10.3109/10715762.2010.507670

[37] Nguyen T, Nioi P, Pickett CB. The Nrf2-antioxidant response element signaling pathway and its activation by oxidative stress. The Journal of Biological Chemistry. 2009;**284**:13291-13295. DOI: 10.1074/jbc.R900010200

[38] Du L, Wang J, Chen Y, Li X, Wang L, Li Y, et al. Novel biphenyl diester derivative AB-38b inhibits NLRP3 inflammasome through Nrf2 activation in diabetic nephropathy. Cell Biology and Toxicology. 2020;**36**:243-260. DOI: 10.1007/s10565-019-09501-8

[39] Sun Y, Yang T, Leak K. Preventive and protective roles of dietary Nrf2 activators against central nervous system

diseases. CNS & Neurological Disorders Drug Targets. 2017;**16**:326-338. DOI: 10.2174/1871527316666170102120211

[40] Scannevin RH, Chollate S, et al. Fumarates promote cytoprotection of central nervous system cells against oxidative stress via the nuclear factor (erythroid-derived 2)-like 2 pathway. Journal of Pharmacology. 2012;**341**:274-284

[41] Lee JM, Li J, Johnson DA, Stein TD, Kraft AD, Calkins MJ, et al. Nrf2, a multi-organ protector? The FASEB Journal. 2005;**19**:1061-1066. DOI: 10.1096/fj.04-2591hyp

[42] Heurtaux T, Bouvier DS, Benani A, Helgueta Romero S, Frauenknecht KB, Mittelbronn M, et al. Normal and pathological NRF2 signalling in the central nervous system. Antioxidants. 2022;**11**:1426. DOI: 10.3390/antiox11081426

[43] Yang Y, Jiang S, Yan J, Li Y, Xin Z, Lin Y, et al. An overview of the molecular mechanisms and novel roles of Nrf2 in neurodegenerative disorders. Cytokine & Growth Factor Reviews. 2015;**26**:47-57. DOI: 10.1016/j.cytogfr.2014.09.002

[44] Tavakkoli A, Iranshahi M, Hasheminezhad SH, Hayes AW, Karimi G. The neuroprotective activities of natural products through the Nrf2 upregulation. Phytotherapy Research. 2019;**33**:2256-2273. DOI: 10.1002/ptr.6427

[45] Zhang M, An C, Gao Y, Leak RK, Chen J, Zhang F. Emerging roles of Nrf2 and phase II antioxidant enzymes in neuroprotection. Progress in Neurobiology. 2013;**100**:30-47. DOI: 10.1016/j.pneurobio.2012.09.003

[46] Wang J, Shen X, Liu J, Chen W, Wu F, Wu W, et al. High glucose mediates NLRP3 inflammasome activation via upregulation of ELF3 expression. Cell Death & Disease. 2020;**11**:1-14. DOI: 10.1038/s41419-020-2598-6

[47] Hung WL, Ho CT, Pan MH. Targeting the NLRP3 inflammasome in neuroinflammation: Health promoting effects of dietary phytochemicals in neurological disorders. Molecular Nutrition & Food Research. 2020;**64**:e1900550. DOI: 10.1002/mnfr.201900550

[48] Hu H, Zhu T, Gong L, Zhao Y, Shao Y, Li S, et al. Transient receptor potential melastatin 2 contributes to neuroinflammation and negatively regulates cognitive outcomes in a pilocarpine-induced mouse model of epilepsy. International Immunopharmacology. 2020;**87**:106824. DOI: 10.1016/j.intimp.2020.106824

[49] Rong S, Wan D, Fan Y, Liu S, Sun K, Huo J, et al. Amentoflavone affects epileptogenesis and exerts neuroprotective effects by inhibiting NLRP3 inflammasome. Frontiers in Pharmacology. 2019;**10**:856. DOI: 10.3389/fphar.2019.00856

[50] Kanneganti T-D. Central roles of NLRs and inflammasomes in viral infection. Nature Reviews. Immunology. 2010;**10**:688-698. DOI: 10.1038/nri2851

[51] Kitazawa M, Cheng D, Tsukamoto MR, Koike MA, Wes PD, Vasilevko V, et al. Blocking IL-1 signaling rescues cognition, attenuates tau pathology, and restores neuronal β-catenin pathway function in an Alzheimer's disease model. Journal of Immunology. 2011;**187**:6539-6549. DOI: 10.4049/jimmunol.1100620

[52] Lu J, Wu D-m, Zheng Y-l, Hu B, Zhang Z-f, Ye Q, et al. Ursolic acid attenuates D-galactose-induced

inflammatory response in mouse prefrontal cortex through inhibiting AGEs/RAGE/NF-κB pathway activation. Cerebral Cortex. 2010;**20**:2540-2548. DOI: 10.1093/cercor/bhq002

[53] Cui X, Zuo P, Zhang Q, Li X, Hu Y, Long J, et al. Chronic systemic D-galactose exposure induces memory loss, neurodegeneration, and oxidative damage in mice: Protective effects of R-α-lipoic acid. Journal of Neuroscience Research. 2006;**83**:1584-1590. DOI: 10.1002/jnr.20899

[54] Wang H, Lv J, Jiang N, Huang H, Wang Q, Liu X. Ginsenoside Re protects against chronic restraint stress-induced cognitive deficits through regulation of NLRP3 and Nrf2 pathways in mice. Phytotherapy Research. 2021;**35**:2523-2535. DOI: 10.1002/ptr.6947

[55] Kobayashi EH, Suzuki T, Funayama R, Nagashima T, Hayashi M, Sekine H, et al. Nrf2suppressesmacrophageinflammatory response by blocking proinflammatory cytokine transcription. Nature Communications. 2016;7:1-14. DOI: 10.1038/ncomms11624

[56] Mills EL, Ryan DG, Prag HA, Dikovskaya D, Menon D, Zaslona Z, et al. Itaconate is an anti-inflammatory metabolite that activates Nrf2 via alkylation of KEAP1. Nature. 2018;**556**:113-117. DOI: 10.1038/nature25986

[57] Yeh W-J, Yang H-Y, Pai M-H, Wu C-H, Chen J-R. Long-term administration of advanced glycation end-product stimulates the activation of NLRP3 inflammasome and sparking the development of renal injury. The Journal of Nutritional Biochemistry. 2017;**39**:68-76. DOI: 10.1016/j.jnutbio.2016.09.014

[58] Song Y, Wang Y, Zhang Y, Geng W, Liu W, Gao Y, et al. Advanced glycation end products regulate anabolic and catabolic activities via NLRP3-inflammasome activation in human nucleus pulposus cells. Journal of Cellular and Molecular Medicine. 2017;**21**:1373-1387. DOI: 10.1111/jcmm.13067

[59] Juliana C, Fernandes-Alnemri T, Kang S, Farias A, Qin F, Alnemri ES. Non-transcriptional priming and deubiquitination regulate NLRP3 inflammasome activation. The Journal of Biological Chemistry. 2012;**287**:36617-36622. DOI: 10.1074/jbc.M112.407130

[60] Li Y-N, Guo Y, Xi M-M, Yang P, Zhou X-Y, Yin S, et al. Saponins from Aralia taibaiensis attenuate D-galactose-induced aging in rats by activating FOXO3a and Nrf2 pathways. Oxidative Medicine and Cellular Longevity. 2014;**2014**:320513. DOI: 10.1155/2014/320513

[61] Lv J, Su W, Yu Q, Zhang M, Di C, Lin X, et al. Heme oxygenase-1 protects airway epithelium against apoptosis by targeting the proinflammatory NLRP3–RXR axis in asthma. The Journal of Biological Chemistry. 2018;**293**:18454-18465. DOI: 10.1074/jbc.RA118.004950

[62] Kim S-J, Lee S-M. NLRP3 inflammasome activation in D-galactosamine and lipopolysaccharide-induced acute liver failure: Role of heme oxygenase-1. Free Radical Biology & Medicine. 2013;**65**:997-1004. DOI: 10.1016/j.freeradbiomed.2013.08.178

[63] Wang K, Lv Q, Miao Y-m, Qiao S-m, Dai Y, Wei Z-f. Cardamonin, a natural flavone, alleviates inflammatory bowel disease by the inhibition of NLRP3 inflammasome activation via an AhR/Nrf2/NQO1 pathway. Biochemical Pharmacology. 2018;**155**:494-509. DOI: 10.1016/j.bcp.2018.07.039

[64] Schröder M. Endoplasmic reticulum stress responses. Cellular and Molecular

Life Sciences. 2008;**65**:862-894. DOI: 10.1007/s00018-007-7383-5

[65] Tengholm A, Hellman B, Gylfe E. The endoplasmic reticulum is a glucose-modulated high-affinity sink for Ca2+ in mouse pancreatic β-cells. The Journal of Physiology. 2001;**530**:533. DOI: 10.1111/j.1469-7793.2001.0533k.x

[66] Guo Y, Zhang C, Wang C, Huang Y, Liu J, Chu H, et al. Thioredoxin-1 is a target to attenuate alzheimer-like pathology in diabetic encephalopathy by alleviating endoplasmic reticulum stress and oxidative stress. Frontiers in Physiology. 2021;**12**:651105. DOI: 10.3389/fphys.2021.651105

[67] Gregersen N, Bross P. Protein misfolding and cellular stress: An overview. Methods in Molecular Biology. 2010;**648**:3-23. DOI: 10.1007/978-1-60761-756-3_1

[68] Ye J, Han Y, Chen X, Xie J, Liu X, Qiao S, et al. L-carnitine attenuates H2O2-induced neuron apoptosis via inhibition of endoplasmic reticulum stress. Neurochemistry International. 2014;**78**:86-95. DOI: 10.1016/j.neuint.2014.08.009

[69] Cai M et al. The molecular mechanisms of excessive hippocampal endoplasmic reticulum stress depressing cognition-related proteins expression and the regulatory effects of Nrf2. Neuroscience. 2020;**431**:152-165. DOI: 10.1016/j.neuroscience.2020.02.001

[70] Nakagawa T, Zhu H, Morishima N, Li E, Xu J, Yankner BA, et al. Caspase-12 mediates endoplasmic-reticulum-specific apoptosis and cytotoxicity by amyloid-β. Nature. 2000;**403**:98-103. DOI: 10.1038/47513

[71] Oyadomari S, Mori M. Roles of CHOP/GADD153 in endoplasmic reticulum stress. Cell Death and Differentiation. 2004;**11**:381-389. DOI: 10.1038/sj.cdd.4401373

[72] Ye T, Meng X, Zhai Y, Xie W, Wang R, Sun G, et al. Gastrodin ameliorates cognitive dysfunction in diabetes rat model via the suppression of endoplasmic reticulum stress and NLRP3 inflammasome activation. Frontiers in Pharmacology. 2018;**9**:1346. DOI: 10.3389/fphar.2018.01346

[73] Zhu Y-F, Li X-H, Yuan Z-P, Li C-Y, Tian R-B, Jia W, et al. Allicin improves endoplasmic reticulum stress-related cognitive deficits via PERK/Nrf2 antioxidative signaling pathway. European Journal of Pharmacology. 2015;**762**:239-246. DOI: 10.1016/j.ejphar.2015.06.002

[74] Cullinan SB, Diehl JA. PERK-dependent activation of Nrf2 contributes to redox homeostasis and cell survival following endoplasmic reticulum stress. The Journal of Biological Chemistry. 2004;**279**:20108-20117. DOI: 10.1074/jbc.M314219200

[75] Rom S, Zuluaga-Ramirez V, Gajghate S, Seliga A, Winfield M, Heldt NA, et al. Hyperglycemia-driven neuroinflammation compromises BBB leading to memory loss in both diabetes mellitus (DM) type 1 and type 2 mouse models. Molecular Neurobiology. 2019;**56**:1883-1896. DOI: 10.1007/s12035-018-1195-5

[76] Bogush M, Heldt NA, Persidsky Y. Blood brain barrier injury in diabetes: Unrecognized effects on brain and cognition. Journal of Neuroimmune Pharmacology. 2017;**12**:593-601. DOI: 10.1007/s11481-017-9752-7

[77] Prasad S, Sajja RK, Naik P, Cucullo L. Diabetes mellitus and blood-brain barrier dysfunction: An overview.

Journal of Pharmacovigil. 2014;**2**:125. DOI: 10.4172/2329-6887.1000125

[78] Van Dyken P, Lacoste B. Impact of metabolic syndrome on neuroinflammation and the blood–brain barrier. Frontiers in Neuroscience. 2018;**12**:930. DOI: 10.3389/fnins.2018.00930

[79] Sajja RK, Prasad S, Tang S, Kaisar MA, Cucullo L. Blood-brain barrier disruption in diabetic mice is linked to Nrf2 signaling deficits: Role of ABCB10? Neuroscience Letters. 2017;**653**:152-158. DOI: 10.1016/j.neulet.2017.05.059

[80] Chen B, Lu Y, Chen Y, Cheng J. The role of Nrf2 in oxidative stress-induced endothelial injuries. The Journal of Endocrinology. 2015;**225**:R83-R99. DOI: 10.1530/JOE-14-0662

[81] Yang M-y, Fan Z, Zhang Z, Fan J. MitoQ protects against high glucose-induced brain microvascular endothelial cells injury via the Nrf2/HO-1 pathway. Journal of Pharmacological Sciences. 2021;**145**:105-114. DOI: 10.1016/j.jphs.2020.10.007

[82] Esteras N, Dinkova-Kostova AT, Abramov AY. Nrf2 activation in the treatment of neurodegenerative diseases: A focus on its role in mitochondrial bioenergetics and function. Biological Chemistry. 2016;**397**:383-400. DOI: 10.1515/hsz-2015-0295

[83] Fernandes V, Choudhary M, Kumar A, Singh SB. Proteotoxicity and mitochondrial dynamics in aging diabetic brain. Pharmacological Research. 2020;**159**:104948. DOI: 10.1016/j.phrs.2020.104948

[84] Luo J-S, Ning J-Q, Chen Z-Y, Li W-J, Zhou R-L, Yan R-Y, et al. The role of mitochondrial quality control in cognitive dysfunction in diabetes. Neurochemical Research. 2022;**8**:2158-2172. DOI: 10.1007/s11064-022-03631-y

[85] Burchell VS, Gandhi S, Deas E, Wood NW, Abramov AY, Plun-Favreau H. Targeting mitochondrial dysfunction in neurodegenerative disease: Part I. Expert Opinion on Therapeutic Targets. 2010;**14**:369-385. DOI: 10.1517/14728221003652489

[86] Biswas S, Chida AS, Rahman I. Redox modifications of protein–thiols: Emerging roles in cell signaling. Biochemical Pharmacology. 2006;**71**:551-564. DOI: 10.1016/j.bcp.2005.10.044

[87] Dodson M, Redmann M, Rajasekaran NS, Darley-Usmar V, Zhang J. KEAP1–NRF2 signalling and autophagy in protection against oxidative and reductive proteotoxicity. The Biochemical Journal. 2015;**469**:347-355. DOI: 10.1042/BJ20150568

[88] Tebay LE, Robertson H, Durant ST, Vitale SR, Penning TM, Dinkova-Kostova AT, et al. Mechanisms of activation of the transcription factor Nrf2 by redox stressors, nutrient cues, and energy status and the pathways through which it attenuates degenerative disease. Free Radical Biology & Medicine. 2015;**88**:108-146. DOI: 10.1016/j.freeradbiomed.2015.06.021

[89] Chen H, Tang X, Zhou B, Zhou Z, Xu N, Wang Y. A ROS-mediated mitochondrial pathway and Nrf2 pathway activation are involved in BDE-47 induced apoptosis in Neuro-2a cells. Chemosphere. 2017;**184**:679-686. DOI: 10.1016/j.chemosphere.2017.06.006

[90] Gao J, Liu S, Xu F, Liu Y, Lv C, Deng Y, et al. Trilobatin protects against oxidative injury in neuronal PC12 cells through regulating mitochondrial ROS

homeostasis mediated by AMPK/Nrf2/Sirt3 signaling pathway. Frontiers in Molecular Neuroscience. 2018;**11**:267. DOI: 10.3389/fnmol.2018.00267

[91] Ebrahimi R, Sepand MR, Seyednejad SA, Omidi A, Akbariani M, Gholami M, et al. Ellagic acid reduces methotrexate-induced apoptosis and mitochondrial dysfunction via up-regulating Nrf2 expression and inhibiting the IκBα/NFκB in rats. Daru. 2019;**27**:721-733. DOI: 10.1007/s40199-019-00309-9

[92] Chorley BN, Campbell MR, Wang X, Karaca M, Sambandan D, Bangura F, et al. Identification of novel NRF2-regulated genes by ChIP-Seq: Influence on retinoid X receptor alpha. Nucleic Acids Research. 2012;**40**:7416-7429. DOI: 10.1093/nar/gks409

[93] Kawamura T, Umemura T, Hotta N. Cognitive impairment in diabetic patients: Can diabetic control prevent cognitive decline? Journal of Diabetes Investigation. 2012;**3**:413-423. DOI: 10.1111/j.2040-1124.2012.00234.x

[94] Polvani S, Tarocchi M, Galli A. PPARγ and oxidative stress: Con (β) catenating NRF2 and FOXO. PPAR Research. 2012;**2012**:641087. DOI: 10.1155/2012/641087

[95] Sharma V, Kaur A, Singh TG. Counteracting role of nuclear factor erythroid 2-related factor 2 pathway in Alzheimer's disease. Biomedicine & Pharmacotherapy. 2020;**129**:110373. DOI: 10.1016/j.biopha.2020.110373

[96] Mannan A, Garg N, Singh TG, Kang HK. Peroxisome proliferator-activated receptor-gamma (PPAR-ɣ): Molecular effects and its importance as a novel therapeutic target for cerebral ischemic injury. Neurochemical Research. 2021;**46**:2800-2831. DOI: 10.1007/s11064-021-03402-1

[97] Ahmadian M, Suh JM, Hah N, Liddle C, Atkins AR, Downes M, et al. PPARγ signaling and metabolism: The good, the bad and the future. Nature Medicine. 2013;**19**:557-566. DOI: 10.1038/nm.3159

[98] Cai W, Yang T, Liu H, Han L, Zhang K, Hu X, et al. Peroxisome proliferator-activated receptor γ (PPARγ): A master gatekeeper in CNS injury and repair. Progress in Neurobiology. 2018;**163**:27-58. DOI: 10.1016/j.pneurobio.2017.10.002

[99] Thapa K, Khan H, Sharma U, Grewal AK, Singh TG. Poly (ADP-ribose) polymerase-1 as a promising drug target for neurodegenerative diseases. Life Sciences. 2021;**267**:118975. DOI: 10.1016/j.lfs.2020.118975

[100] Pu D, Zhao Y, Chen J, Lv A, Zhu S, Luo C, et al. Protective effects of sulforaphane on cognitive impairments and AD-like lesions in diabetic mice are associated with the upregulation of Nrf2 transcription activity. Neuroscience. 2018;**381**:35-45. DOI: 10.1016/j.neuroscience.2018.04.017

[101] Wang G, Fang H, Zhen Y, Xu G, Tian J, Zhang Y, et al. Sulforaphane prevents neuronal apoptosis and memory impairment in diabetic rats. Cellular Physiology and Biochemistry. 2016;**39**:901-907. DOI: 10.1159/000447799

[102] Zhang S, Yuan L, Zhang L, Li C, Li J. Prophylactic use of troxerutin can delay the development of diabetic cognitive dysfunction and improve the expression of Nrf2 in the hippocampus on STZ diabetic rats. Behavioural Neurology. 2018;**2018**:8678539. DOI: 10.1155/2018/8678539

[103] Gao M, Kang Y, Zhang L, Li H, Qu C, Luan X, et al. Troxerutin attenuates cognitive decline in the

hippocampus of male diabetic rats by inhibiting NADPH oxidase and activating the Nrf2/ARE signaling pathway. International Journal of Molecular Medicine. 2020;**46**:1239-1248. DOI: 10.3892/ijmm.2020.4653

[104] Zhang L, Ma Q, Zhou Y. Strawberry leaf extract treatment alleviates cognitive impairment by activating Nrf2/HO-1 signaling in rats with streptozotocin-induced diabetes. Frontiers in Aging Neuroscience. 2020;**12**:201. DOI: 10.3389/fnagi.2020.00201

[105] Ma C, Long H. Protective effect of betulin on cognitive decline in streptozotocin (STZ)-induced diabetic rats. Neurotoxicology. 2016;**57**:104-111

[106] Baluchnejadmojarad T, Kiasalari Z, Afshin-Majd S, Ghasemi Z, Roghani M. S-allyl cysteine ameliorates cognitive deficits in streptozotocin-diabetic rats via suppression of oxidative stress, inflammation, and acetylcholinesterase. European Journal of Pharmacology. 2017;**794**:69-76. DOI: 10.1016/j.ejphar.2016.11.033

[107] Luo Z, Wan Q, Han Y, Li Z, Li B. CAPE-pNO2 ameliorates diabetic brain injury through modulating Alzheimer's disease key proteins, oxidation, inflammation and autophagy via a Nrf2-dependent pathway. Life Sciences. 2021;**287**:119929

[108] Chen Y-J, Tang Z-Z, Du L, Liu Y, Lu Q, Ma T-F, et al. A novel compound AB-38b improves diabetes-associated cognitive decline in mice via activation of Nrf2/ARE pathway. Brain Research Bulletin. 2019;**150**:160-167

[109] Wang G, Zhang X, Lu X, Liu J, Zhang Z, Wei Z, et al. Fish oil supplementation attenuates cognitive impairment by inhibiting neuroinflammation in STZ-induced diabetic rats. Aging (Albany NY). 2020;**12**:15281-15289. DOI: 10.18632/aging.103426

[110] Ke B, Zhang T, An T, Lu R. Soy isoflavones ameliorate the cognitive dysfunction of Goto-Kakizaki rats by activating the Nrf2-HO-1 signalling pathway. Aging (Albany NY). 2020;**12**:21344-21354. DOI: 10.18632/aging.103877

[111] Wang X, Fang H, Xu G, Yang Y, Xu R, Liu Q, et al. Resveratrol prevents cognitive impairment in type 2 diabetic mice by upregulating Nrf2 expression and transcriptional level. Diabetes Metabolosm Syndrome in Obesity. 2020;**13**:1061-1075. DOI: 10.2147/DMSO.S243560

[112] Sadi G, Konat D. Resveratrol regulates oxidative biomarkers and antioxidant enzymes in the brain of streptozotocin-induced diabetic rats. Pharmaceutical Biology. 2016;**54**:1156-1163. DOI: 10.3109/13880209.2015

[113] Li H, Qin T, Li M, Ma S. Thymol improves high-fat diet-induced cognitive deficits in mice via ameliorating brain insulin resistance and upregulating NRF2/HO-1 pathway. Metabolic Brain Disease. 2017;**32**:385-393. DOI: 10.1007/s11011-016-9921-z

[114] Feng Y, Chu A, Luo Q, Wu M, Shi X, Chen Y. The protective effect of astaxanthin on cognitive function via inhibition of oxidative stress and inflammation in the brains of chronic T2DM rats. Frontiers in Pharmacology. 2018;**9**:748. DOI: 10.3389/fphar.2018.00748

[115] Zhai Y, Meng X, Luo Y, Wu Y, Ye T, Zhou P, et al. Notoginsenoside R1 ameliorates diabetic encephalopathy by activating the Nrf2 pathway and inhibiting NLRP3 inflammasome

activation. Oncotarget. 2018;**9**:9344-9363. DOI: 10.18632/oncotarget.24295

[116] Liu Y-W, Liu X-L, Kong L, Zhang M-Y, Chen Y-J, Zhu X, et al. Neuroprotection of quercetin on central neurons against chronic high glucose through enhancement of Nrf2/ARE/glyoxalase-1 pathway mediated by phosphorylation regulation. Biomedicine & Pharmacotherapy. 2019;**109**:2145-2154. DOI: 10.1016/j.biopha.2018.11.066

[117] Singh NK, Garabadu D. Quercetin exhibits α7nAChR/Nrf2/HO-1-mediated neuroprotection against STZ-induced mitochondrial toxicity and cognitive impairments in experimental rodents. Neurotoxicity Research. 2021;**39**:1859-1879. DOI: 10.1007/s12640-021-00410-5

[118] Ghawi SK, Methven L, Niranjan K. The potential to intensify sulforaphane formation in cooked broccoli (Brassica oleracea var. italica) using mustard seeds (Sinapis alba). Food Chemistry. 2013;**138**:1734-1741. DOI: 10.1016/j.foodchem.2012.10.119

[119] Hwang E-S, Jeffery EH. Induction of quinone reductase by sulforaphane and sulforaphane N-acetylcysteine conjugate in murine hepatoma cells. Journal of Medicinal Food. 2005;**8**:198-203. DOI: 10.1089/jmf.2005.8.198

[120] McNaughton S, Marks G. Development of a food composition database for the estimation of dietary intakes of glucosinolates, the biologically active constituents of cruciferous vegetables. The British Journal of Nutrition. 2003;**90**:687-697. DOI: 10.1079/bjn2003917

[121] Sones K, Heaney RK, Fenwick GR. An estimate of the mean daily intake of glucosinolates from cruciferous vegetables in the UK. Journal of the Science of Food and Agriculture. 1984;**35**:712-720

[122] Dash PK, Zhao J, Orsi SA, Zhang M, Moore AN. Sulforaphane improves cognitive function administered following traumatic brain injury. Neuroscience Letters. 2009;**460**:103-107. DOI: 10.1016/j.neulet.2009.04.028

[123] Sun Y, Yang T, Mao L, Zhang F. Sulforaphane protects against brain diseases: Roles of cytoprotective enzymes. Austin Journal of Cerebrovascular Diseases. 2017:4

[124] Su Z-Y, Zhang C, Lee JH, Shu L, Wu T-Y, Khor TO, et al. Requirement and epigenetics reprogramming of Nrf2 in suppression of tumor promoter TPA-induced mouse skin cell transformation by SulforaphaneActivating Nrf2 blocks TPA-induced JB6 transformation by SFN. Cancer Prevention Research (Philadelphia, Pa.). 2014;7:319-329. DOI: 10.1158/1940-6207.CAPR-13-0313-T

[125] Hu R, Xu C, Shen G, Jain MR, Khor TO, Gopalkrishnan A, et al. Gene expression profiles induced by cancer chemopreventive isothiocyanate sulforaphane in the liver of C57BL/6J mice and C57BL/6J/Nrf2 (−/−) mice. Cancer Letters. 2006;**243**:170-192. DOI: 10.1016/j.canlet.2005.11.050

[126] Hu C, Eggler AL, Mesecar AD, Van Breemen RB. Modification of keap1 cysteine residues by sulforaphane. Chemical Research in Toxicology. 2011;**24**:515-521. DOI: 10.1021/tx100389r

[127] Hong F, Freeman ML, Liebler DC. Identification of sensor cysteines in human Keap1 modified by the cancer chemopreventive agent sulforaphane. Chemical Research in Toxicology. 2005;**18**:1917-1926. DOI: 10.1021/tx0502138

[128] Shang G, Tang X, Gao P, Guo F, Liu H, Zhao Z, et al. Sulforaphane attenuation of experimental diabetic

nephropathy involves GSK-3 beta/Fyn/Nrf2 signaling pathway. The Journal of Nutritional Biochemistry. 2015;**26**:596-606. DOI: 10.1016/j.jnutbio.2014.12.008

[129] Bergström P, Andersson HC, Gao Y, Karlsson J-O, Nodin C, Anderson MF, et al. Repeated transient sulforaphane stimulation in astrocytes leads to prolonged Nrf2-mediated gene expression and protection from superoxide-induced damage. Neuropharmacology. 2011;**60**:343-353. DOI: 10.1016/j.neuropharm.2010.09.023

[130] Uddin MS, Al Mamun A, Jakaria M, Thangapandiyan S, Ahmad J, Rahman MA, et al. Emerging promise of sulforaphane-mediated Nrf2 signaling cascade against neurological disorders. Science Total Environment. 2020;**707**:135624. DOI: 10.1016/j.scitotenv.2019.135624

[131] Ashrafizadeh M, Ahmadi Z, Yaribeygi H, Sathyapalan T, Sahebkar A. Astaxanthin and Nrf2 signaling pathway: A novel target for new therapeutic approaches. Mini Reviews in Medicinal Chemistry. 2022;**22**:312-321. DOI: 10.2174/1389557521666210505112834

[132] Chen Y, Tang J, Zhang Y, Du J, Wang Y, Yu H, et al. Astaxanthin alleviates gestational diabetes mellitus in mice through suppression of oxidative stress. Naunyn-Schmiedeberg's Archives of Pharmacology. 2020;**393**:2517-2527. DOI: 10.1007/s00210-020-01861-x

[133] Jeong SM, Kim Y-J. Astaxanthin treatment induces maturation and functional change of myeloid-derived suppressor cells in tumor-bearing mice. Antioxidants. 2020;**9**:350. DOI: 10.3390/antiox9040350

[134] Mohammadi S, Barzegari A, Dehnad A, Barar J, Omidi Y. Astaxanthin protects mesenchymal stem cells from oxidative stress by direct scavenging of free radicals and modulation of cell signaling. Chemico-Biological Interactions. 2021;**333**:109324. DOI: 10.1016/j.cbi.2020.109324

[135] Yuan L, Qu Y, Li Q, An T, Chen Z, Chen Y, et al. Protective effect of astaxanthin against La2O3 nanoparticles induced neurotoxicity by activating PI3K/AKT/Nrf-2 signaling in mice. Food and Chemical Toxicology. 2020;**144**:111582. DOI: 10.1016/j.fct.2020.111582

[136] Landon R, Gueguen V, Petite H, Letourneur D, Pavon-Djavid G, Anagnostou F. Impact of astaxanthin on diabetes pathogenesis and chronic complications. Marine Drugs. 2020;**18**:357. DOI: 10.3390/md18070357

[137] Xu L, Zhu J, Yin W, Ding X. Astaxanthin improves cognitive deficits from oxidative stress, nitric oxide synthase and inflammation through upregulation of PI3K/Akt in diabetes rat. International Journal of Clinical Pathology. 2015;**8**:6083

[138] Huang D-D, Shi G, Jiang Y, Yao C, Zhu C. A review on the potential of resveratrol in prevention and therapy of diabetes and diabetic complications. Biomedicine & Pharmacotherapy. 2020;**125**:109767. DOI: 10.1016/j.biopha.2019.109767

[139] Pan MH, Wu JC, Ho CT, Lai CS. Antiobesity molecular mechanisms of action: Resveratrol and pterostilbene. BioFactors. 2018;**44**:50-60. DOI: 10.1002/biof.1409

[140] Yang DK, Kang H-S. Anti-diabetic effect of cotreatment with quercetin and resveratrol in streptozotocin-induced diabetic rats. Biomolecular Therapy (Seoul). 2018;**26**:130. DOI: 10.4062/biomolther.2017.254

[141] Elshaer M, Chen Y, Wang XJ, Tang X. Resveratrol: An overview of its anti-cancer mechanisms. Life Sciences. 2018;**207**:340-349. DOI: 10.1016/j.lfs.2018.06.028

[142] De La Lastra CA, Villegas I. Resveratrol as an anti-inflammatory and anti-aging agent: Mechanisms and clinical implications. Molecular Nutrition & Food Research. 2005;**49**:405-430. DOI: 10.1002/mnfr.200500022

[143] Seyyedebrahimi S, Khodabandehloo H, Nasli Esfahani E, Meshkani R. The effects of resveratrol on markers of oxidative stress in patients with type 2 diabetes: A randomized, double-blind, placebo-controlled clinical trial. Acta Diabetologica. 2018;**55**:341-353. DOI: 10.1007/s00592-017-1098-3

[144] Ebrahimpour S, Zakeri M, Esmaeili A. Crosstalk between obesity, diabetes, and alzheimer's disease: Introducing quercetin as an effective triple herbal medicine. Ageing Research Reviews. 2020;**62**:101095. DOI: 10.1016/j.arr.2020.101095

[145] Sun GY, Chen Z, Jasmer KJ, Chuang DY, Gu Z, Hannink M, et al. Quercetin attenuates inflammatory responses in BV-2 microglial cells: Role of MAPKs on the Nrf2 pathway and induction of heme oxygenase-1. PLoS One. 2015;**10**:e0141509. DOI: 10.1371/journal.pone.0141509

[146] Wang D, Zhao J, Li S, Shen G, Hu S. Quercetin attenuates domoic acid-induced cognitive deficits in mice. Nutritional Neuroscience. 2018;**21**:123-131. DOI: 10.1080/1028415X.2016.1231438

[147] Yano S, Tokumitsu H, Soderling TR. Calcium promotes cell survival through CaM-K kinase activation of the protein-kinase-B pathway. Nature. 1998;**396**:584-587. DOI: 10.1038/25147

[148] Hajduch E, Litherland GJ, Hundal HS. Protein kinase B (PKB/Akt)–a key regulator of glucose transport? FEBS Letters. 2001;**492**:199-203. DOI: 10.1016/s0014-5793(01)02242-6

Chapter 6

Oxidative Stress in Substance Use Disorders: Endogenous and Exogenous Mechanisms of Repair

Pedro Lirio, Matheus Santos, Igor Silva and Livia Rodrigues

Abstract

Substance use disorders (SUDs) can be defined as patterns of symptoms caused by the abusive consumption of recreational or prescribed substances that an individual continues to use despite their negative effects. Oxidative stress is one of the main pathophysiological processes occasioned by SUDs in different brain areas. Oxidative damage and subsequent deleterious symptoms can happen because of the consumption of psychoactive drugs, both stimulants and depressants. This chapter focuses on SUDs associated with depressant drugs, such as alcohol, opioids, benzodiazepines, and their effects on the central nervous system (CNS). We present the main characteristics of the SUDs and later explore endogenous mechanisms of repair, such as neuroglia and the endocannabinoid system. We also examine the neuroprotective effects of exogenous substances such as phytocannabinoids (e.g., cannabidiol) and N-acetylcysteine (NAC), which have shown important roles in anti-inflammatory pathways and antioxidative cascades, and how these molecules can be potential tools in the treatment of neurological symptoms of SUDs.

Keywords: oxidative stress, substance use disorder, depressant drugs, N-acetylcysteine, phytocannabinoids

1. Introduction

Oxidative stress is a concept introduced in the 1980s and is related to the imbalance or insufficiency of antioxidant systems to detriment of the production of reactive oxygen species (ROS) (**Table 1**), causing significant biological damage [1]. The production of ROS arises endogenously and exogenously. Cellular respiration is performed by mitochondria, in which there is a massive production of superoxide anion ($O2^-$), (being one of the main endogenous sources of ROS) [2]. Activation of pro-inflammatory cytokines/chemokines exacerbated activation of the hypothalamic–pituitary–adrenal axis, and consequent mobilization of mineralocorticoid receptors are other examples of endogenous forms of ROS production. Furthermore, the environment, pollution, and drug consumption, such as alcohol and opioids, can exacerbate the process of free radicals' synthesis [3].

ROS	Radicals	Non-Radicals
	•O_2^-	H_2O_2
	•OH	HOCl
	ROO•	
	HCO_3•	$ONOO^-$
	HO_2•	

Table 1.
Main ROS produced in CNS.

The central nervous system (CNS) is a highly susceptible site for oxidative stress process to occur once it demands a high consumption of O2 and nutrients, such as glycose and lactate. CNS contains a high content of fatty acids and lipid-rich structures, which also leads to increased oxidative stress [4]. High concentration of ROS can lead to increased blood–brain barrier permeability, decrease of synaptogenesis and neuroplasticity and can cause mitochondrial dysfunction, due to the processes of lipid peroxidation and protein oxidation caused by the binding of free radicals to primordial functional structures of the CNS [5].

To balance the production of ROS and free radicals in the CNS there are endogenous mechanisms of repair, such as the glutathione antioxidant system, which consists of the most abundant non-enzymatic antioxidant system (mainly responsible for the metabolization of peroxides and inactivation of free radicals) and some others, like neuroglia and the endocannabinoid system, which both will be discussed later in this chapter [6]. Alongside the endogenous mechanisms of repair, some exogenous molecules can be useful to reduce oxidative stress and subsequent damage. N-acetylcysteine or simply NAC plays a role in glutathione synthesis and shows prominent antioxidant properties. In addition, it has presented important therapeutic potential in the management of drug abuse [7, 8]. Moreover, recent studies have shown that phytocannabinoids, such as cannabidiol and Δ9-tetrahydrocannabinol (THC), presented antioxidant-like properties when involved in the regulation and balance of both prevention and recovery of damage caused by ROS and free radicals [9].

In this chapter, we will discuss about substance use disorders of depressant drugs, such as benzodiazepines, opioids, and alcohol, highlighting endogenous and exogenous mechanisms of repair and prevention of oxidative stress damage.

2. Substance use disorders: depressant drugs

2.1 Alcohol use disorder

Alcohol (ethanol) is one of the most consumed drugs in the world. Its consumption is related to the causes of 6–9% of all neuronal, mental, and substance use disorders [10]. Alcohol use disorder (AUD) is defined in DSM-V as a pattern of alcohol consumption that corresponds to two or more of 11 characteristics that an individual presented during a period of 12 months. These criteria can also be used for any other substance use disorder (SUD) [11].

In the U.S. alone, at least one-third of the population is likely to meet the criteria for AUD at some point in their lives [12]. It is well known that abusive ethanol consumption, including episodic binge drinking and more continuous pattern of drinking, can cause impairment in neuronal and cognitive aspects of individuals. Binge drinking, characterized by a heavy, fast, and episodic alcohol intake, is very popular among younger age groups. The prevalence of adolescents (15 to 19 years old) that consume alcohol in a binge pattern is similar to the overall population, with an almost 1:1 ratio in the Americas, Europe, and Western Pacific [13]. Likewise, binge drinking also presents potential damage to the brain, especially for adolescents and young adults, given that many areas of their brains are immature and still very plastic [14]. Although the impairments caused by AUD are explicit and abusive consumption of alcohol is a concern worldwide, studies are still needed to elucidate the consequences of alcohol consumption in the central nervous system (CNS).

So far, the action of ethanol in the brain is seemingly general and interacts with different areas and neuronal circuits, modulating the release of different neurotransmitters, such as GABA, glutamate, and dopamine. Nonetheless, alcohol has a widespread effect on the CNS; GABA receptors are the only ones known to have a binding site to ethanol, where it works as a positive allosteric modulator (**Figure 1**). This characteristic of the drug makes it even harder to specify its effects and mechanisms of action [16]. The higher levels of oxidative stress indicate a higher chance of developing memory impairments in several modalities, such as working, spatial and recognition memory; increase in anxiety levels, that can contribute to the onset of generalized anxiety disorder; reduced attention and decision-making skills, which can lead to future SUDs as well as impairment in social skills, even though alcohol is considered a social drug. Damage in the central nervous system caused by AUD is mostly regarded as an interrelation between oxidative stress and neuroinflammation promoted by neuroglia when microglia and astrocytes are activated.

The activation of microglia and astrocytes through neuroinflammatory process are modulated by the release of pro- and anti-inflammatory cytokines and other inflammatory mediators that also participate in the oxidative balance of the CNS. Microglial activation caused by abusive ethanol consumption raises the levels of pro-inflammatory cytokines, such as TNF-α, IL-1, and IL-6, as well as inflammatory mediators COX-2 and nitric oxide [15, 17–19]. The pro-inflammatory response to alcohol is mediated by toll-like receptors (TLR), in which the TLR-4 has an important role in both microglia and astrocytes. Given that acute ethanol effects on GABAergic signaling are mediated by TLR-4 [20], alcohol chronic consumption impairs the immune response through TLR-4 and can also lead to the development of neurological disorders, such as Alzheimer's disease, demonstrating its key role in ethanol-related neurodegeneration [14, 21, 22]. Adolescent TLR-4 knockout mice did not raise alcohol preference over time when compared to control, as well as ethanol-induced levels of BDNF and Fos B in medial prefrontal cortex [23]. Thus, AUD appears to maintain and even escalate the ethanol intake, through the increase of glial inflammatory signaling, which in turn causes neurodegeneration and possible neuronal disorders [24].

Alongside the innate antioxidant and anti-inflammatory mechanisms, promoted by antioxidant enzymes, cytokines, and the glial cells, some exogenous and other endogenous substances are currently investigated as an alternative to neutralize neuronal damage caused by alcohol-induced oxidative stress. The endocannabinoid system (ECS) covers a vast area of the CNS and participates in the modulation of microglia inflammatory response, granting both phyto- and endocannabinoids an important role in alcohol-induced oxidative balance and neurodegeneration. Cannabinoid receptor type 1 (CB1) activation can lead to a reduction of

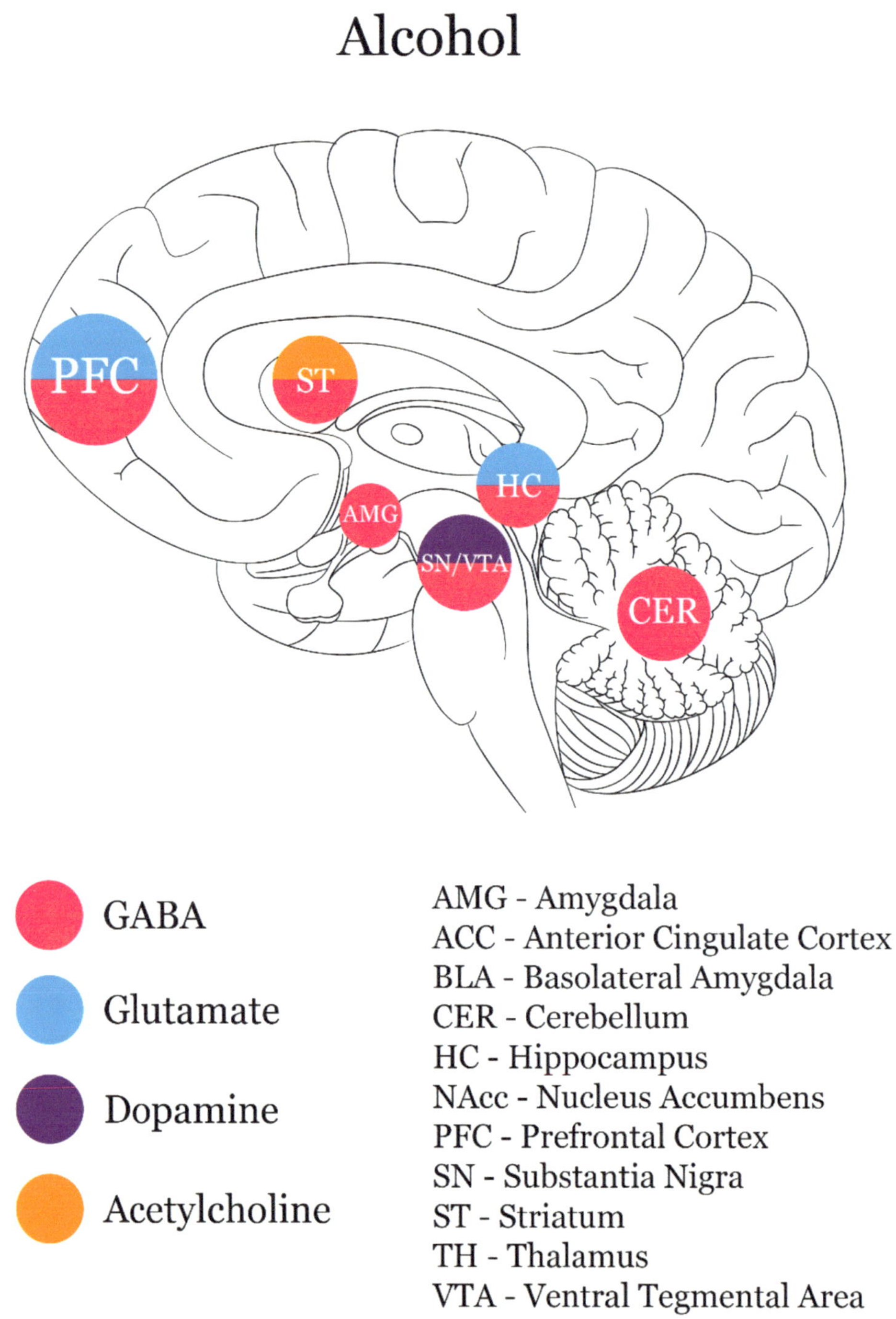

Figure 1.
Anatomical distribution of the main neurotransmitter systems that alcohol interacts with, with special attention to areas related to the symptoms presented in AUD. Ethanol, as a depressant drug, interacts with GABAergic synapses across the brain, whereas it also interferes with excitatory glutamatergic neurotransmission in the PFC and HC, altering synaptic and extra-synaptic glutamate transport [15]. In ST and the neuroanatomical complex SN/VTA, monoamines, such as dopamine and acetylcholine, and their receptors are also affected by alcohol consumption.

pro-inflammatory cytokines, iNOS, ROS, and higher protection from excitotoxicity, despite it being most expressed in neurons compared to glial cells [25–28]. Cannabinoid receptor type 2 (CB2) is more expressed in microglial cells. Thus, it is intrinsically involved in the regulation of the inflammatory response. Activation of CB2 is related to a decrease in the expression of pro-inflammatory cytokines, such as TNF-α, INF-γ, IL-1, IL-2, IL-6 and IL-12, iNOS, and chemokines [29–31]. It has been demonstrated that a chronic model of binge drinking in adolescent rats lead to an increase in levels of the enzymes that synthesize two important endocannabinoids (eCBs): anandamide (AEA) and 2-arachidonoylglycerol (2-AG), as well as pro-inflammatory mediators, such as TLR4, TNF-, COX-2 and GFAP in the medial prefrontal cortex (mPFC) [32].

The metabolic modulation of AEA (and other N-Acetylethanolamines) by URB597, a FAAH selective inhibitor, prevented the production of ROS in both acute and chronic binge drinking models in adolescent rats [33]. Pretreatment with URB597 was able to prevent an increase of INF-γ and TNF-α levels in prefrontal cortex (PFC) and hippocampus in a chronic binge model in adolescent rats, as well as reduced IL-4, IL-10, and BDNF expression in PFC [34]. AEA also interacts with PPAR receptors, which are well known to modulate anti-inflammatory and antioxidant responses through the upregulation of NRF2, a transcriptional factor responsible to maintain redox homeostasis that inhibits oxidative stress and neuroinflammation [35]. Thus far, URB597 show to be a very promising neuroprotective substance against alcohol-induced oxidative stress and neuroinflammation.

N-acetylcysteine (NAC) is a promising, cheap, and accessible neuroprotective substance against oxidative damage caused by many SUDs, including AUD. In rodent models, NAC prevented the increase of pro-inflammatory and the decrease of anti-inflammatory cytokines in frontal cortex and hippocampus, and reduced leptin and corticosterone levels and hypoactive behavior after alcohol cessation in a chronic treatment [36, 37]. It also reduced motivation, seeking-behavior, and reacquisition in a model of ethanol self-administration [38]. NAC also reduced the anxiety-like behavior and oxidative stress levels in a withdrawal period after chronic ethanol exposure in zebrafish, with similar results obtained after acute exposure [7, 8]. Coadministration of NAC and acetylsalicylic acid provided some promising results on ethanol chronic intake and relapse, as this combination reduced both behaviors in rats. The synergism between these substances may act on the oxidative stress and neuroinflammatory cycle through different mechanisms, suggesting these cycles might contribute to relapse episodes and chronic alcohol consumption [39, 40].

2.2 Opioid use disorder

Opioid use disorder (OUD) refers to a problematic pattern of opioid use that affects 2 million people aged 12 or older in the United States, according to the 2018 national survey on drug use and health [41] that meet the DSM-5 criteria for OUD, while other 10 million misuse opioids—illegal or medically prescribed—in some degree [42].

OUD when regarded as the “opioid epidemic” is considered to be one of the most severe public health crisis in US history [38, 39], comprising at least four waves throughout the last decades, which have their own particularities when it comes to social, demographic, and health-related contexts [40, 41]. Over-prescribing licit pain relievers and low prices of street heroin and illicit fentanyl are examples of factors influencing the outspread consumption of opioids and subsequent OUD [42].

Diagnosis and evaluation of the severity of OUD in a given patient are assessed by the display of aberrant behaviors included as criteria in the DSM-5. Mild OUD is diagnosed by the presence of 2–3 criteria, while 4–5 criteria are considered moderate, and 6 or more are severe on the OUD spectrum [43].

Opiates, endogenous or exogenous, can act on three main receptors, but morphine and other alkaloid opioids have a high affinity for the receptor mu (μ), which is widely distributed in the brain. Receptor mu binding is responsible for the psychotropic effects of opioids, and eventual respiratory depression and death by overdose. Even though the vast majority of deaths as a result of illicit drug consumption are attributed to opioid use [44], it is important to notice that prolonged use of opioids and non-fatal overdoses are clinically significant, as they can lead to neurocognitive impairments [45, 46]. Oxidative stress has been demonstrated to be one of the main mechanisms of neuronal damage induced by licit and illicit opioids, as shown by studies in both animal and human subjects.

In rodent models, heroin can induce oxidative damage in the DNA, proteins, and membrane lipids in the brains of mice [47], while in rats, the repeated administration of licit pain reliever morphine is known to induce blood–brain barrier (BBB) disruptions during the withdrawal phase [48, 49] and this damage appears to be accentuated by oxidative stress [50]. Oxidative markers are also altered in the brains of Wistar rats exposed to codeine, another common pharmacologically relevant opioid [51].

One common trait in the incidence of brain oxidative stress after exposure to opioids (**Figure 2**) is that not only oxidative damage is noticed, but some endogenous antioxidant mechanisms are also impaired. The glutathione antioxidant system seems to be one of the most affected, as a decrease in both brain intracellular reduced glutathione (GSH) levels and glutathione peroxidase (GSH-Px) activity has been observed in mice [52]. In humans, post-mortem studies seem to suggest a similar pro-oxidant pathway, as reduced glutathione has been found decreased in the frontal, temporal, parietal and occipital cortex, brain stem, hippocampus, and white matter of deceased individuals with OUD that consumed heroin [53]. In mice brain, not only glutathione peroxidase activity was found impaired after heroin exposure, but also many other classic enzymatic antioxidant mechanisms, such as superoxide dismutase and catalase [54].

Therefore, secondary endogenous mechanisms of neuroprotection and repair are of great importance, as they can be potential targets for therapies aiming to manage OUD and subsequent oxidative damage caused by it. The endocannabinoid system has a well-known interaction with the opioid receptor system regarding anesthesia [55] and its role in the modulation of reward neural circuits have placed it as one of the most promising candidates for the non-opioid management of OUD [56, 57]. However, it is still not clear whether the endocannabinoid system could be further explored to counterbalance the deleterious pro-oxidative effects of OUD.

However, exogenous substances like NAC have demonstrated a lot of potential in treating OUD-associated oxidative damage. NAC is hydrolyzed after entering the cell and releases cysteine, a GSH precursor [58, 59]. As GSH levels are frequently impaired by OUD-induced oxidative stress, NAC appears as a frontrunner as a mitigator of ROS formation, as a consequence of opioid toxicity. Although the literature lacks more information regarding NAC-induced central neuroprotection against oxidative stress caused by OUD, in the periphery, NAC is capable of attenuating oxidative stress in the liver of tramadol-treated rats by stimulating the production of GSH [60].

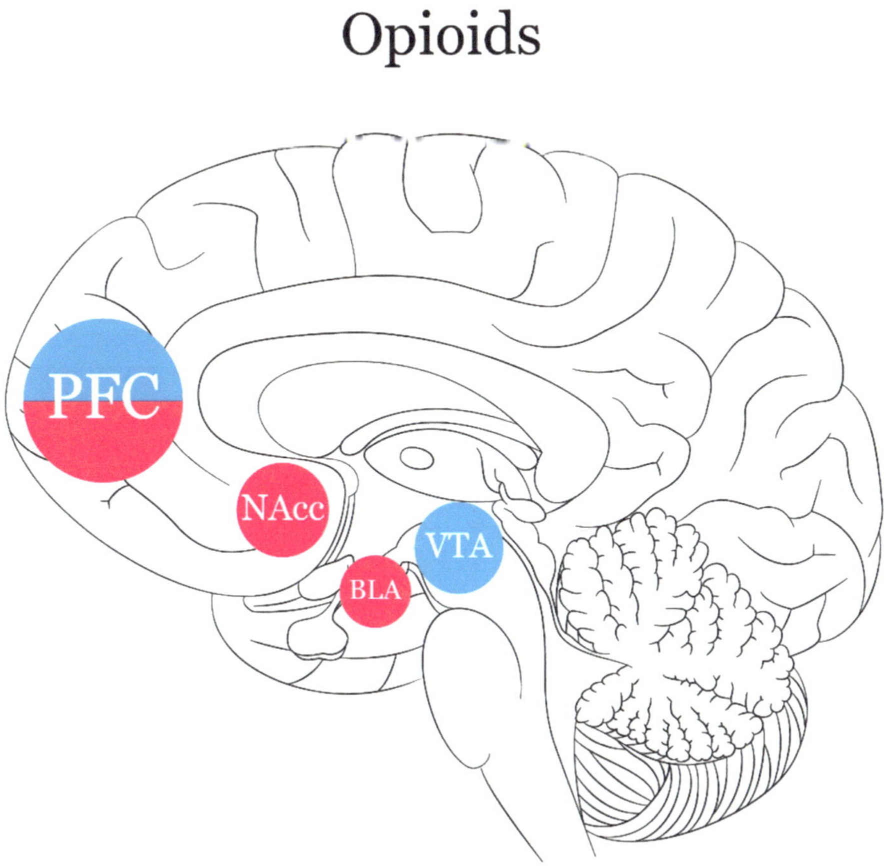

Figure 2.
Anatomical distribution of the main neurotransmitter systems that opioids interact with, with special attention to areas related to the symptoms presented in opioid SUD. While opioids have their own receptors in the CNS through which they exert their effects, they can also interact with both GABAergic and glutamatergic synapses, especially in PFC, but also in NAcc, BLA, and VTA.

2.3 Benzodiazepine use disorder

Benzodiazepines (BDZ) have been widely used anxiolytics since their development in the 1960s, comprising several drugs with anticonvulsant and sedative-hypnotic properties employed in various types of anxiety-related disorders. However, long-term treatment with BDZs is demonstrated to induce physiological dependence even in therapeutic doses, leading to increased anxiety and insomnia as a result of the withdrawal from the drug after months of use.

Therefore, there is a concern that most prescribed BDZs can potentially lead to drug abuse by former medical users and even nonmedical and recreational use by self-medication. In the United States, 6 million citizens in an age range starting from 12 years and older misused tranquilizers in 2016, making this class of drug the third most commonly misused illicit substance following marijuana (15%), prescription opioids (4.1%) and with numbers similar to those of cocaine abuse, comprising approximately 2.2% of the population [61].

Consequently, it is possible to subcategorize BDZ use disorder into two patterns of abuse: deliberate abuse and unintentional abuse [62]. Deliberate abuse happens when individuals misuse BDZs in order to achieve an altered state of mind due to the drug's psychoactive effects. Other SUDs often accompany this type of use, and abusers might even take BDZs to self-medicate the withdrawal symptoms of other drugs. On the other hand, unintentional abuse is characterized by a prescribed BDZ medical use that is taken out of its original therapeutical purpose. Those individuals often overuse BDZ by taking a higher dose than necessary due to developed pharmacological tolerance, by using a BDZ drug other than the prescribed one, or might continue taking the medication after the intended treatment to self-medicate episodic anxiety symptoms or even withdrawal symptoms caused by BDZs themselves.

In addition to this, it is fundamental to take into consideration the long-lasting effects that BDZ use disorder might have on brain physiology. When it comes to damage caused by oxidative stress, however, it appears that this pathological process is not present in BDZ use disorder, in both deliberate and unintentional abuse. BDZs bind allosterically to GABAA receptors in several brain areas (**Figure 3**), potentially reducing excitotoxicity induced by glutamate release and subsequent oxidative damage that arises from this. On the other hand, the pharmacological interactions between BDZs and other drugs are frequent in deliberate abuse and should be taken into consideration when evaluating the extent of the neurotoxic effects of both drugs [63].

It is notably dangerous relationship between BDZ and opioids, especially the unintentional abuse of BDZ in concomitancy with OUD [64], as these drug classes

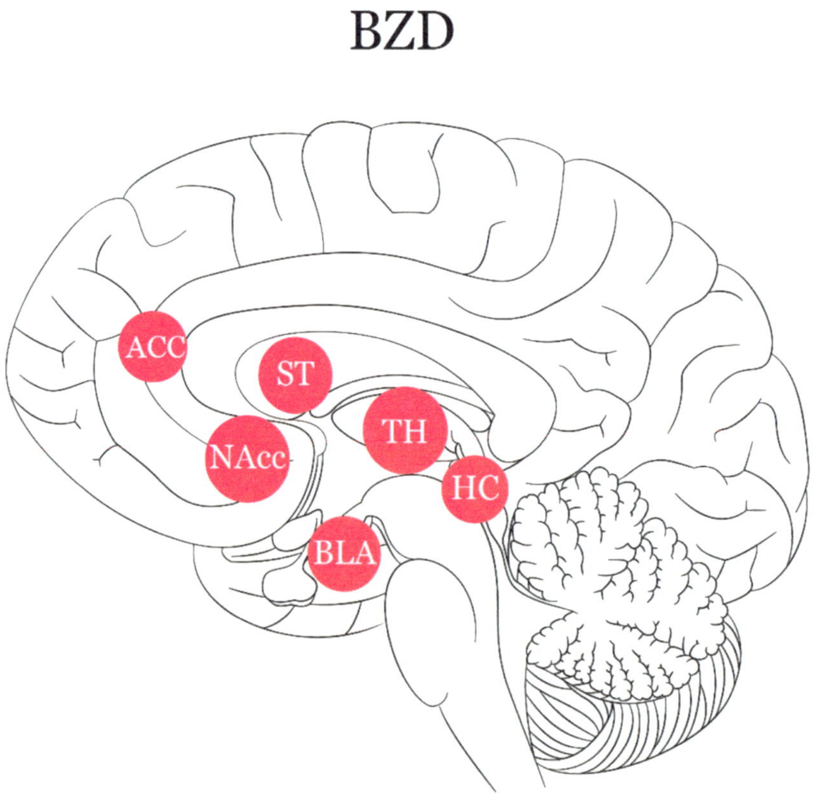

Figure 3.
Anatomical distribution of the main neurotransmitter systems that BZDs interact with, with special attention to areas related to the symptoms presented in benzodiazepine SUD. Given the pharmacological nature of such drugs, adverse effects on BZD abuse are mediated almost entirely by GABAergic transmission, mainly in subcortical areas.

are commonly prescribed together. Between 2002 and 2014, the proportion of opioid recipients prescribed concomitantly with a BDZ drug each year had a relative increase of 41% [65], and this pharmacological association is estimated to increase risk of an emergency room visit or inpatient admission for opioid overdose [66].

3. Conclusions

In summary, oxidative stress seems to be involved in alterations in neurochemistry of CNS, neuroinflammatory conditions, and cognitive dysfunctions related to excessive consumption of different depressant drugs, such as benzodiazepines, opioids, and ethanol. Endogenous substances, such as endocannabinoids and exogenous as N-acetylcisteine or phytocannabinoids, have been highlighted as potential drugs/mechanisms of prevention and repair of oxidative stress damage in CNS. Future perspectives include the more selective tests that can point to the mechanisms through which exogenous protective substances act on CNS and interact with endogenous mechanisms of repair on each specific SUD.

Acknowledgements

The authors would like to acknowledge FAPES and CAPES for providing scholarship funding to three of the co-authors involved in writing this chapter.

Conflict of interest

The authors declare no conflict of interest.

Thanks

The authors also would like to thank Radael Rezende Rodrigues, Jr. for the help in the making of the figures.

Appendices and nomenclature

ROS	reactive species of oxygen
CNS	central nervous system
TL4	toll-like receptor 4
SUD	substance use disorder
AUD	alcohol use disorder
OUD	opioid use disorder
NAC	N-acetylcisteine
BZD	benzodiazepine
CBD	cannabidiol
THC	Δ9-tetrahydrocannabinol
COX2	cyclooxygenase 2
IL-1/6	interleukin 1/6

TNFα	tumor necrosis factor α
GABA	gamma aminobutyric Acid
GSH	glutathione
GSH-Px	glutathione Peroxidase
AMG	amygdala
ACC	anterior Cingulate Cortex
BLA	basolateral Amygdala
CER	cerebellum
HC	hippocampus
NAcc	nucleus accumbens
PFC	prefrontal cortex
SN	substantia nigra
ST	striatum
TH	thalamus
VTA	ventral tegmental area

Author details

Pedro Lirio, Matheus Santos, Igor Silva and Livia Rodrigues*
Laboratory of Neurotoxicology and Psycopharmacology, Department of Physiological Sciences, Federal University of Espirito Santo, Vitória, Brazil

*Address all correspondence to: susuvi@gmail.com

References

[1] Sies H. Oxidative stress: A concept in redox biology and medicine. Redox Biology. 2015;**4**:180-183. DOI: 10.1016/j.redox.2015.01.002

[2] Salim S. Oxidative stress and the central nervous system. The Journal of Pharmacology and Experimental Therapeutics. 2017;**360**(1):201-205. DOI: 10.1124/jpet.116.237503

[3] Patel M. Targeting oxidative stress in central nervous system disorders. Trends in Pharmacological Sciences. 2016;**36**(9):768-778. DOI: 10.1016/j.tips.2016.06.007

[4] Lehtinen MK, Bonni A. Modeling oxidative stress in the central nervous system. Current Molecular Medicine. 2006;**6**(8):871-881. DOI: 10.2174/156652406779010786

[5] Filomeni G, De Zio D, Cecconi F. Oxidative stress and autophagy: The clash between damage and metabolic needs. Cell Death and Differentiation. 2015;**22**(3):377-388. DOI: 10.1038/cdd.2014.150

[6] Chen Y, Qin C, Huang J, et al. The role of astrocytes in oxidative stress of central nervous system: A mixed blessing. Cell Proliferation. 2020;**53**(3):e12781. DOI: 10.1111/cpr.12781

[7] Mocelin R, Marcon M, D'ambros S, Herrmann AP, da Rosa Araujo AS, Piato A. Behavioral and biochemical effects of N-Acetylcysteine in zebrafish acutely exposed to ethanol. Neurochemical Research. 2018;**43**(2):458-464. DOI: 10.1007/s11064-017-2442-2

[8] Mocelin R, Marcon M, da Rosa Araujo AS, Herrmann AP, Piato A. Withdrawal effects following repeated ethanol exposure are prevented by N-acetylcysteine in zebrafish. Progress in Neuro-Psychopharmacology & Biological Psychiatry. 2019;**93**:161-170. DOI: 10.1016/j.pnpbp.2019.03.014

[9] García-Baos A, Alegre-Zurano L, Cantacorps L, Martín-Sánchez A, Valverde O. Role of cannabinoids in alcohol-induced neuroinflammation. Progress in Neuro-Psychopharmacology & Biological Psychiatry. 2021;**104**: 110054. DOI: 10.1016/j.pnpbp.2020.110054

[10] Patel V, Chisholm D, Parikh R, Charlson FJ, Degenhardt L, Dua T, et al. Addressing the burden of mental, neurological, and substance use disorders: Key messages from disease control priorities, 3rd edition. Lancet 2016;**387**(10028):1672-1685

[11] APA. Diagnostic and Statistical Manual of Mental Disorders. 5th ed. Arlington, VA, USA: APA; 2013

[12] Grant BF, Goldstein RB, Saha TD, Chou SP, Jung J, Zhang H, et al. Epidemiology of DSM-5 alcohol use disorder: Results from the National Epidemiologic Survey on alcohol and related conditions III. JAMA Psychiatry. 2015;**72**(8):757-766. DOI: 10.1001/jamapsychiatry.2015.0584

[13] World Health Organization. Global Status Report on Alcohol and Health 2018. Switzerland: World Health Organization; 2018. pp. 1-472

[14] Floreani NA, Rump TJ, Muneer PMA, Alikunju S, Morsey BM, Brodie MR, et al. Alcohol-induced interactive phosphorylation of Src and toll-like receptor regulates the secretion

of inflammatory mediators by human astrocytes. Journal of Neuroimmune Pharmacology. 2010;**5**(4):533-545. DOI: 10.1007/s11481-010-9213-z

[15] Berríos-Cárcamo P, Quezada M, Quintanilla ME, Morales P, Ezquer M, Herrera-Marschitz M, et al. Oxidative stress and Neuroinflammation as a pivot in drug abuse. A focus on the therapeutic potential of antioxidant and anti-inflammatory agents and biomolecules. Antioxidants. 2020;**9**(9):830. DOI: 10.3390/antiox9090830

[16] Pascual M, Montesinos J, Guerri C. Role of the innate immune system in the neuropathological consequences induced by adolescent binge drinking. Journal of Neuroscience Research. 2017;**96**(5):765-780

[17] McClain JA, Morris SA, Deeny MA, Marshall SA, Hayes DM, Kiser ZM, et al. Adolescent binge alcohol exposure induces long-lasting partial activation of microglia. Brain, Behavior, and Immunity. 2011;**25**:S120-S128

[18] Vallés SL, Blanco AM, Pascual M, Guerri C. Chronic ethanol treatment enhances inflammatory mediators and cell death in the brain and in astrocytes. Brain Pathology. 2004;**14**(4):365-371

[19] Bajo M, Madamba SG, Roberto M, Blednov YA, Sagi VN, Roberts E, et al. Innate immune factors modulate ethanol interaction with GABAergic transmission in mouse central amygdala. Brain, Behavior, and Immunity. 2014;**40**:191-202

[20] Pla A, Pascual M, Guerri C. Autophagy constitutes a protective mechanism against ethanol toxicity in mouse astrocytes and neurons. PLoS One. 2016;**11**(4):e0153097. DOI: 10.1371/journal.pone.0153097

[21] Walter S, Letiembre M, Liu Y, Heine H, Penke B, Hao W, et al. Role of the toll-like receptor 4 in neuroinflammation in Alzheimer's disease. Cellular Physiology and Biochemistry. 2007;**20**(6):947-956

[22] Montesinos J, Pascual M, Rodríguez-Arias M, Miñarro J, Guerri C. Involvement of TLR4 in the long-term epigenetic changes, rewarding and anxiety effects induced by intermittent ethanol treatment in adolescence. Brain, Behavior, and Immunity. 2016;**53**:159-171

[23] Adermark L, Bowers MS. Disentangling the role of astrocytes in alcohol use disorder. Alcoholism, Clinical and Experimental Research. 2016;**40**(9):1802-1816

[24] Lou Z-Y, Yu W-B, Chen J, Li L, Jiang L-S, Xiao B-G, et al. Neuroprotective effect is driven through the upregulation of CB1 receptor in experimental autoimmune encephalomyelitis. Journal of Molecular Neuroscience. 2016;**58**(2):193-200

[25] Ribeiro R, Wen J, Li S, Zhang Y. Involvement of ERK1/2, cPLA2 and NF-κB in microglia suppression by cannabinoid receptor agonists and antagonists. Prostaglandins & Other Lipid Mediators. 2013;**100**:1-14

[26] Stella N. Cannabinoid and cannabinoid-like receptors in microglia, astrocytes, and astrocytomas. Glia. 2010;**58**(9):1017-1030

[27] Croxford JL, Yamamura T. Cannabinoids and the immune system: Potential for the treatment of inflammatory diseases? Journal of Neuroimmunology. 2005;**166**(1-2):3-18

[28] Pelição R, Santos MC, Freitas-lima LC, Meyrelles SS, Vasquez EC,

Nakamura-palacios EM, et al. URB597 inhibits oxidative stress induced by alcohol binging in the prefrontal cortex of adolescent rats. Neuroscience Letters. 2016;**624**:17-22. DOI: 10.1016/j.neulet.2016.04.068

[29] Bellozi PMQ, Pelição R, Santos MC, Lima IVA, Saliba SW, Vieira ÉLM, et al. URB597 ameliorates the deleterious effects induced by binge alcohol consumption in adolescent rats. Neuroscience Letters. 2019;**711**(July):134408. DOI: 10.1016/j.neulet.2019.134408

[30] Zhang M, An C, Gao Y, Leak RK, Chen J, Zhang F. Emerging roles of Nrf2 and phase II antioxidant enzymes in neuroprotection. Progress in Neurobiology. 2013;**100**:30-47

[31] Cai W, Yang T, Liu H, Han L, Zhang K, Hu X, et al. Peroxisome proliferator-activated receptor γ (PPARγ): A master gatekeeper in CNS injury and repair. Progress in Neurobiology. 2018;**163**:27-58

[32] Rojo AI, Pajares M, García-Yagüe AJ, Buendia I, Van Leuven F, Yamamoto M, et al. Deficiency in the transcription factor NRF2 worsens inflammatory parameters in a mouse model with combined tauopathy and amyloidopathy. Redox Biology. 2018;**18**:173-180

[33] Schneider R, Santos CF, Clarimundo V, Dalmaz C, Elisabetsky E, Gomez R. N-acetylcysteine prevents behavioral and biochemical changes induced by alcohol cessation in rats. Alcohol. 2015;**49**(3):259-263. DOI: 10.1016/j.alcohol.2015.01.009

[34] Schneider R, Bandiera S, Souza DG, Bellaver B, Caletti G, Quincozes-Santos A, et al. N-acetylcysteine prevents alcohol related Neuroinflammation in rats. Neurochemical Research. 2017;**0**(0):1-7

[35] Lebourgeois S, González-Marín MC, Antol J, Naassila M, Vilpoux C. Evaluation of N-acetylcysteine on ethanol self-administration in ethanol-dependent rats. Neuropharmacology. 2019;**150**:112-120. DOI: 10.1016/j.neuropharm.2019.03.010

[36] Israel Y, Quintanilla ME, Ezquer F, Morales P, Santapau D, Berríos-Cárcamo P, et al. Aspirin and N-acetylcysteine co-administration markedly inhibit chronic ethanol intake and block relapse binge drinking: Role of neuroinflammation-oxidative stress self-perpetuation. Addiction Biology. 2021;**26**(1):1-13

[37] Quintanilla ME, Ezquer F, Morales P, Ezquer M, Olivares B, Santapau D, et al. N-Acetylcysteine and acetylsalicylic acid inhibit alcohol consumption by different mechanisms: Combined protection. Frontiers in Behavioral Neuroscience. 2020;**14**(July):1-15

[38] Substance Abuse and Mental Health Services Administration. Key Substance Use and Mental Health Indicators in the United States. Results from the 2018 National Survey on Drug Use and Health. Rockville, MD, USA: Substance Abuse and Mental Health Services Administration; 2019

[39] Ahrnsbrak R, Bose J, Hedden SL, et al. Key Substance Use and Mental Health Indicators in the United States: Results from the 2016 National Survey on Drug Use and Health (HHS Publication No. SMA 17-5044, NSDUH Series H-52). Rockville, MD: Center for Behavioral Health Statistics and Quality, Substance Abuse and Mental Health Services Administration; 2017

[40] Vowles Kevin E, Mc Entee Mindy L, Siyahhana JP, Tessaa F, Ney John P, van der Goes David N. Rates of opioid misuse, abuse, and addiction in

chronic pain. Pain. 2015;**156**(4):569-576. DOI: 10.1097/01.j.pain.0000460357.01998.f1

[41] Kosten TR, Baxter LE. Review article: Effective management of opioid withdrawal symptoms: A gateway to opioid dependence treatment. The American Journal on Addictions. 2019;**28**(2):55-62. DOI: 10.1111/ajad.12862

[42] Volkow ND, Blanco C. The changing opioid crisis: Development, challenges and opportunities. Molecular Psychiatry. 2021;**26**(1):218-233. DOI: 10.1038/s41380-020-0661-4

[43] Ciccarone D. The triple wave epidemic: Supply and demand drivers of the US opioid overdose crisis. The International Journal on Drug Policy. 2019;**71**:183-188. DOI: 10.1016/j.drugpo.2019.01.010

[44] World Drug Report. 2020. United Nations publication, Sales No. E.20.XI.6

[45] Winstanley EL, Mahoney JJ 3rd, Castillo F, Comer SD. Neurocognitive impairments and brain abnormalities resulting from opioid-related overdoses: A systematic review. Drug and Alcohol Dependence. 2021;**226**:108838. DOI: 10.1016/j.drugalcdep.2021.108838

[46] Ahmad-Molaei L, Hassanian-Moghaddam H, Farnaghi F, Tomaz C, Haghparast A. Delay-dependent impairments in memory and motor functions after acute methadone overdose in rats. Frontiers in Pharmacology. 2018;**9**:1023. DOI: 10.3389/fphar.2018.01023

[47] Xu B, Wang Z, Li G, Li B, Lin H, Zheng R, et al. Heroin-administered mice involved in oxidative stress and exogenous antioxidant-alleviated withdrawal syndrome. Basic & Clinical Pharmacology & Toxicology. 2006;**99**(2):153-161. DOI: 10.1111/j.1742-7843.2006.pto_461.x

[48] Sharma HS, Ali SF. Alterations in blood-brain barrier function by morphine and methamphetamine. Annals of the New York Academy of Sciences. 2006;**1074**:198-224. DOI: 10.1196/annals.1369.020

[49] Sharma HS, Lundstedt T, Boman A, Lek P, Seifert E, Wiklund L, et al. A potent serotonin-modulating compound AP-267 attenuates morphine withdrawal-induced blood-brain barrier dysfunction in rats. Annals of the New York Academy of Sciences. 2006;**1074**:482-496. DOI: 10.1196/annals.1369.049

[50] Sharma HS, Sjöquist PO, Ali SF. Alterations in blood-brain barrier function and brain pathology by morphine in the rat. Neuroprotective effects of antioxidant H-290/51. Acta Neurochirurgica. Supplement. 2010;**106**:61-66. DOI: 10.1007/978-3-211-98811-4_10

[51] Archibong VB, Ekanem T, Igiri A, Ofutet EO, Ifie JE. The effect of codeine administration on oxidative stress biomarkers and the expression of the neuron-specific enolase in the brain of Wistar rats. Naunyn-Schmiedeberg's Archives of Pharmacology. 2021 Aug;**394**(8):1665-1673. DOI: 10.1007/s00210-021-02094-2

[52] Abdel-Zaher AO, Mostafa MG, Farghaly HS, Hamdy MM, Abdel-Hady RH. Role of oxidative stress and inducible nitric oxide synthase in morphine-induced tolerance and dependence in mice. Effect of alpha-lipoic acid. Behavioural Brain Research. 2013;**15**(247):17-26. DOI: 10.1016/j.bbr.2013.02.034

[53] Gutowicz M, Kaźmierczak B, Barańczyk-Kuźma A. The influence of

heroin abuse on glutathione-dependent enzymes in human brain. Drug and Alcohol Dependence. 2011;**113**(1):8-12. DOI: 10.1016/j.drugalcdep.2010.06.020

[54] Bruehl S, Burns JW, Morgan A, Koltyn K, Gupta R, Buvanendran A, et al. The association between endogenous opioid function and morphine responsiveness: A moderating role for endocannabinoids. Pain. 2019;**160**(3):676-687. DOI: 10.1097/j.pain.0000000000001447

[55] Woodhams SG, Chapman V, Finn DP, Hohmann AG, Neugebauer V. The cannabinoid system and pain. Neuropharmacology. 2017;**124**:105-120. DOI: 10.1016/j.neuropharm.2017.06.015

[56] Fattore L, Deiana S, Spano SM, Cossu G, Fadda P, Scherma M, et al. Endocannabinoid system and opioid addiction: Behavioural aspects. Pharmacology Biochemistry and Behavior. 2005;**81**(2):343-359. DOI: 10.1016/j.pbb.2005.01.031

[57] Alvarez-Jaimes L, Polis I, Parsons LH. Attenuation of cue-induced heroin-seeking behavior by cannabinoid CB1 antagonist infusions into the nucleus accumbens core and prefrontal cortex, but not basolateral amygdala. Neuropsychopharmacology. 2008;**33**(10):2483-2493. DOI: 10.1038/sj.npp.1301630

[58] He XH, Jordan CJ, Vemuri K, Bi GH, Zhan J, Gardner EL, et al. Cannabinoid CB1 receptor neutral antagonist AM4113 inhibits heroin self-administration without depressive side effects in rats. Acta Pharmacologica Sinica. 2019;**40**(3):365-373. DOI: 10.1038/s41401-018-0059-x

[59] Arakawa M, Ito Y. N-acetylcysteine and neurodegenerative diseases: Basic and clinical pharmacology. Cerebellum. 2007;**6**:308-314. DOI: 10.1080/14734220601142878

[60] Tardiolo G, Bramanti P, Mazzon E. Overview on the effects of N-Acetylcysteine in neurodegenerative diseases. Molecules. 2018;**23**(12):3305. DOI: 10.3390/molecules23123305

[61] Adikwu E, Bokolo B. Melatonin and N-Acetylcysteine as remedies for tramadol-induced hepatotoxicity in albino rats. Advanced pharmaceutical bulletin. 2017;**7**(3):367-374. DOI: 10.15171/apb.2017.044

[62] Votaw VR, Geyer R, Rieselbach MM, McHugh RK. The epidemiology of benzodiazepine misuse: A systematic review. Drug and Alcohol Dependence. 2019;**200**:95-114. DOI: 10.1016/j.drugalcdep.2019.02.033

[63] O'brien C. P. Benzodiazepine use, abuse, and dependence. The Journal of Clinical Psychiatry. 2005;**66**(Suppl. 2):28-33

[64] Votaw VR, Witkiewitz K, Valeri L, Bogunovic O, McHugh RK. Nonmedical prescription sedative/tranquilizer use in alcohol and opioid use disorders. Addictive Behaviors. 2019;**88**:48-55. DOI: 10.1016/j.addbeh.2018.08.010

[65] Hwang CS, Kang EM, Kornegay CJ, Staffa JA, Jones CM, McAninch JK. Trends in the concomitant prescribing of opioids and benzodiazepines, 2002-2014. American Journal of Preventive Medicine. 2016;**51**(2):151-160. DOI: 10.1016/j.amepre.2016.02.014

[66] Sun EC, Dixit A, Humphreys K, Darnall BD, Baker LC, Mackey S. Association between concurrent use of prescription opioids and benzodiazepines and overdose: Retrospective analysis. BMJ. 2017;**356**:j760. DOI: 10.1136/bmj.j760

Chapter 7

Oxidative Stress in Cardiovascular Diseases

Laura Mourino-Alvarez, Tamara Sastre-Oliva, Nerea Corbacho-Alonso and Maria G. Barderas

Abstract

Cardiovascular diseases encompass a range of pathologies that affect the heart or blood vessels. Oxidative stress is an important factor that contributes to the development of these pathologies. Adverse effects due to oxidative stress manifest when there is an imbalance between the production and elimination of reactive oxygen species (ROS), or when physiological mechanisms of repair for oxidative injury are overburdened. This chapter focuses on ROS accumulation and antioxidant system deficiencies in the context of their influence on cardiovascular disease. We also discuss the importance of high throughput approaches, such as proteomics, with regard to their role in advancing the field of precision medicine for cardiovascular diseases, while keeping in mind the ultimate goal of improving patient care and quality of life.

Keywords: cardiovascular disease, endothelial dysfunction, reactive oxygen species, thiol compounds, oxidative stress

1. Introduction

Cardiovascular diseases (CVDs) constitute a major cause of global mortality and are an important source of rising healthcare costs; furthermore, they result in significant decreases in the quality of life (QoL) of those who suffer from these conditions [1, 2]. The great majority of CVDs are chronic diseases, and hence, rising incidence translates into higher prevalence, which leads to a healthcare burden that may persist over decades. Prevalence has long been on the rise in almost all countries, with reported increases from 271 million in 1990 to 523 million in 2019, while CVD-associated deaths have also followed suite, increasing from 12.1 million to 18.6 million in the same period [3]. CVDs encompass a range of pathologies that affect the heart or blood vessels, including ischemic heart disease, cerebrovascular disease (stroke), peripheral arterial disease, heart failure, and heart valve disease. Hypercholesterolemia, hypertension, and diabetes, which are associated with oxidative stress (OS) and inflammatory activation, are well-known risk factors for CVD [4–7].

OS originates in cells and tissues when the balance between oxidative and antioxidative compounds (or mechanisms) is disrupted in favor of oxidation. This imbalance between ROS production and antioxidant defenses may be due to increased ROS formation (e.g., superoxide, hydroxyl, nitric oxide, or hydrogen peroxide) and/or insufficiency in antioxidants (e.g., ascorbate, alpha-tocopherol, or thiol-based redox compounds) [8]. The consequences of OS include lipid peroxidation, membrane damage, and the activation of proteases, nucleases, and protein kinases. With respect to vascular functions, excess ROS regulates the release of factors that drive vasoconstriction or vasodilation, leading to endothelial cell (EC) damage, vascular smooth muscle cell (VSMC) hyperplasia, and structural remodeling (**Figure 1**) [9, 10]. Regulation of vascular tone is critical for the maintenance of cardiovascular health. In this sense, OS plays an important role as an initiator of the EC dysfunction. In physiological conditions, maintenance of appropriate endothelial function provides vasorelaxant properties through the release of vasoactive substances. Nevertheless, under OS conditions, there exists an imbalance between the production of vasoprotective and vasorelaxant factors and vasoconstrictor substances by the endothelium [11]. In addition, oxidative stress could induce vascular inflammation and injury through activation of the transcription factors, upregulation of adhesion molecules, stimulation of chemokine production, and recruitment of inflammatory cells [9, 10].

In the following sections, we focus primarily on the OS-related mechanisms that have been associated with CVD development. Firstly, we discuss how ROS accumulate and the importance of this accumulation with regard to endothelial dysfunction and

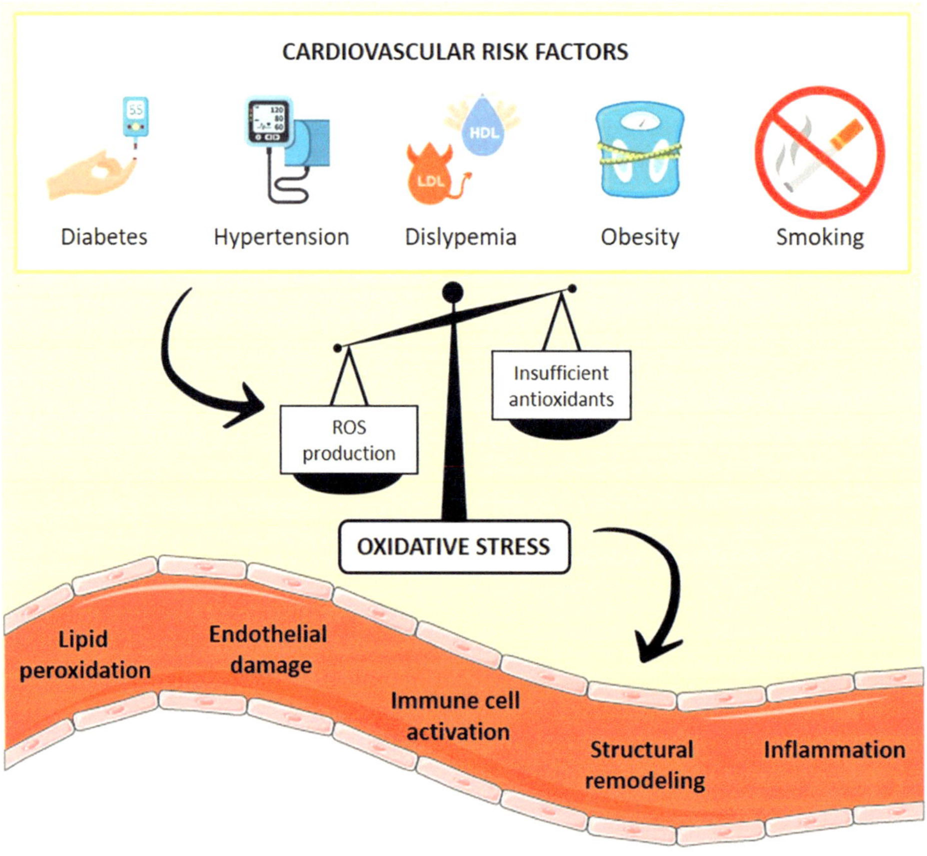

Figure 1.
Cardiovascular risk factors disrupt oxidative metabolism and negatively affect the vascular system through lipid peroxidation, endothelial damage, immune cell activation, structural remodeling, or inflammation.

the vascular system. Next, we concentrate on the role of thiols in the antioxidant system and the consequences of the suppression of this defense system. Finally, we highlight the importance of high throughput approaches, such as proteomics, with respect to their contributions to the advances in the field of precision medicine related to CVD, while keeping in mind the ultimate goal of improving patient care and QoL.

2. ROS accumulation: mitochondrial and ER-related stress

Endothelial damage, vascular dysfunction, cardiac remodeling, immune cell activation, and systemic inflammation are key processes implicated in CVDs, all of which are affected by ROS accumulation [12–15]. The primary oxidase system underlying OS in vascular disease is the NADPH oxidase (NOX) system, the main function of which is to produce ROS [16]. Importantly, while ROS production was originally considered to incite harmful effects, and low levels of ROS are necessary for physiological processes, including cell proliferation, migration, differentiation, and cytoskeletal organization [17]. However, excessive activation of NOX also activates other oxidase systems to maintain OS (**Figure 2**). These secondary mechanisms include, but are not limited to, endothelial nitric oxide synthase (eNOS) uncoupling, mitochondrial stress, and endoplasmic reticulum (ER) stress, which also contribute to redox changes in CVD [18–20]. eNOS is primarily responsible for the generation of vasoprotective nitric oxide (NO). Endothelial-derived NO is an important vasodilator and a potent inhibitor of platelet aggregation and leukocyte adhesion. As such, NO is one of the most important anti-atherogenic factors in vasculature [21]. Under normal conditions, eNOS exists as a dimer that is stabilized by the essential cofactor, tetrahydrobiopterin (BH4). However, when BH4 is inactivated due to excess ROS, the

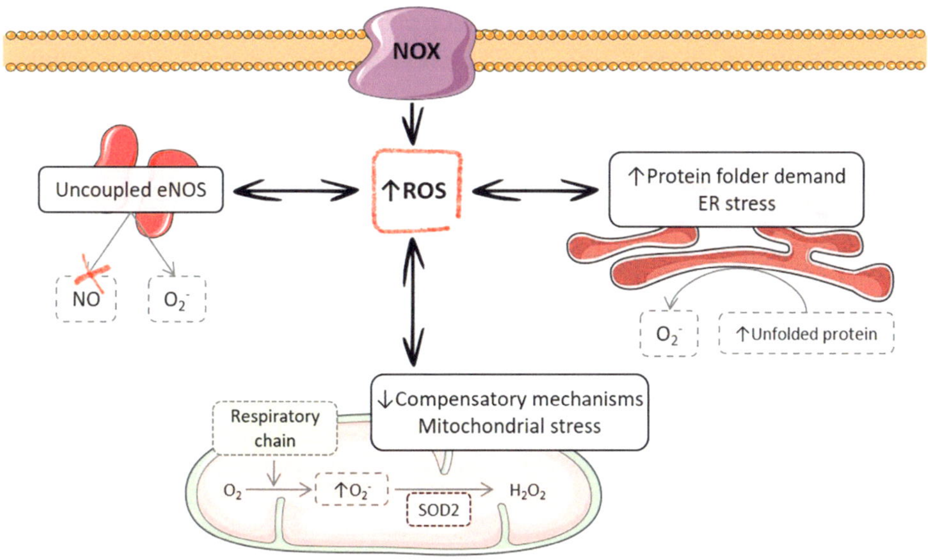

Figure 2.
Excessive activation of NADPH oxidase (NOX) increases the production of ROS to harmful levels. Subsequently, secondary mechanisms, such as endothelial nitric oxide synthase (eNOS) uncoupling, mitochondrial stress, and endoplasmic reticulum (ER) stress, are triggered, perpetuating oxidative stress and injurious effects in a vicious cycle.

dimer breaks down, resulting in the production of vaso-injurious superoxide ($O2^-$) as opposed to NO [22].

The cross-talk between the NOX system and mitochondria may cause ROS-induced release of ROS, a positive feed-forward mechanism of ROS production. As a result, NOX-derived ROS increases the production of mitochondrial ROS, which in turn stimulates NOX activation [23–25]. One consequence of normal mitochondrial metabolism and homeostasis is the production of controlled levels of ROS caused by the reduction of O2 to $O2^-$ through the respiratory chain [26]. Under physiological conditions, these levels are safe and normal cellular activity is maintained through the protection of mitochondria by a multilayer network of antioxidant systems. One particular example is the manganese-dependent superoxide dismutase (SOD2) enzyme, which converts the $O2^-$ anion to hydrogen peroxide (H2O2), which is subsequently transformed to water by protective mechanisms, such as catalase, glutathione peroxidase, and peroxiredoxins [27]. Nevertheless, if ROS levels exceed the normal range, the resulting oxidative environment is harmful and cannot be compensated for by these protective mechanisms [23, 27, 28]. In relation to this, mitochondrial dysfunction also has important implications for CVD. The protective role of different antioxidant systems against atherosclerosis is related to H2O2 metabolisms, such as that driven by the catalase or paraoxonase family, as seen in mouse models [29–31]. Mitochondrial dysfunction is associated with the redox inactivation of the mitochondrial deacetylase Sirtuin 3 (also named Silent acting type information regulation 2 homolog 3, SIRT3) and the mitochondrial antioxidant SOD2 also contributes to the development of hypertension [32, 33]. These discoveries brought forth the use of mitochondrial-targeted interventions as therapeutic agents. Mitoquinone (MitoQ), a mitochondrial ROS scavenger, has been shown to reduce macrophage infiltration into atherosclerotic plaques in a mouse model of metabolic syndrome [34]. Moreover, MitoQ protects against the development of hypertension by improving endothelial NO bioavailability and reduces cardiac hypertrophy in models of hypertension [35]. Similar favorable vascular responses have been demonstrated in humans treated with MitoQ. After oral supplementation, MitoQ was associated with a decrease in plasma oxidized low-density lipoprotein (LDL), a circulating marker of OS, and the amelioration of endothelial function [36].

The ER is increasingly being recognized as an important player in the redox pathophysiology of the cardiovascular system [37–39]. Protein folding contributes to the generation of ROS since ER oxidoreductases serve as terminal electron acceptors with oxygen. When ER stress develops, demands on protein folding are excessively increased and oxygen is not completely reduced, leading to the generation of $O2^-$ anions with the consequent production of H2O2 or other ROS [40, 41]. The ER and mitochondria are closely associated as the protein folding process is energy-dependent. As a consequence, ATP depletion may occur during ER stress, which can stimulate oxidative phosphorylation in the mitochondria due to increased ATP need, subsequently resulting in greater ROS generation. In addition, mitochondrial ROS production will also increase if unfolded proteins accumulate in the ER via the release of Ca^{2+} or interactions with SOD [42, 43]. Of note, ER stress has been shown to play an important role in vascular cell phenotypic switching, de-differentiation, calcification, and apoptosis, contributing to endothelial dysfunction and vascular remodeling in hypertension and in atherosclerosis [44, 45].

With regard to these data, it has been inferred that compounds, which can alleviate ER stress could be pharmacological options to elicit antioxidant effects. One example is the antihypertensive drug guanabenz, which is demonstrated to confer protection against the detrimental accumulation of misfolded proteins in cardiac myocytes [46]. However, there is much controversy about this compound, since it has been reported to induce β-cell dysfunction *in vitro* and *in vivo* (in rodents) and it may lead to impaired glucose tolerance [47]. Other ER stress inhibitors, such as tauroursodeoxycholic acid (TUDCA) or 4-phenylbutyrate (4-PBA), also have beneficial effects on the cardiovascular system. Administration of TUDCA has been shown to alleviate myocardial contractile dysfunction and reduce blood pressure in rodents [48–50]. Likewise, 4-PBA protects against atherosclerotic lesion growth [51], prevents cardiac rupture and remodeling by inhibiting cardiac apoptotic and fibrotic signaling pathways [52], and attenuates interstitial fibrosis and cardiac hypertrophy caused by pressure overload [53]. These studies are among the studies that provide support to the proof of concept that modulation of ER stress can improve cardiovascular function.

3. Thiol-based redox compounds in CVD

Thiol-based redox compounds, such as thioredoxins (Trxs), glutaredoxins (Grxs), and peroxiredoxins (Prxs), are primary contributors to the intracellular redox state, thereby modulating metabolism, signaling, and cell survival pathways [54]. In proteins, the thiol groups of cysteine side chains are highly susceptible to reversible or irreversible oxidative modifications. Besides forming disulfide bonds between two different proteins, these protein thiols can also form disulfide bridges with low-molecular-weight thiols like glutathione, they can be oxidized to sulfenic, sulfinic, and sulfonic, or suffer S-nitrosylation [55]. These modifications can alter the activity of numerous proteins that contain cysteines (Cys) as their anionic form. As such, thiolate (RS^-) plays critical role in protein structure, function, and regulation. Although the Trxs and Grxs systems function differently, they both maintain a reduced intracellular redox state in mammalian cells by reducing protein thiols. Oxidized Grx is reduced by reduced glutathione (GSH), and the oxidized glutathione (GSSG) is then recycled by glutathione reductase (GR) at the expense of NADPH. The Grx system involves the GR, Grx, and the glutathione redox pair (GSH/GSSG). Under oxidative conditions, in which the GSH concentration decreases and GSSG increases, Grx is more likely to be oxidized [55]. Thus, the GSH-to-GSSG ratio (GSH/GSSG) in the cell is an important marker of the redox environment and a major determinant of the cellular redox potential. Meanwhile, Prxs are a family of conserved abundant Cys-based peroxidases that consist of six Prx isoforms (Prx1–6) in mammalian cells. This family plays an essential role in the detoxification of hydrogen peroxide, aliphatic and aromatic hydroperoxides, and peroxynitrite [56, 57].

Currently, the implications of these thiol-based redox compounds on CVDs are being widely studied [55, 58, 59]. Endogenous Trx has the potential to decrease reperfusion-induced arrhythmia [60] and is a central mediator of cardiomyocyte growth [61]. Upregulation of Trxs has been detected in ischemia [62, 63] and cardiac failure [64, 65], probably as a compensatory mechanism in response to myocardial injury. In fact, the role of Trx1, the cytosolic isoform of the Trx family, in myocardial hypertrophy is identified to be fundamental as it acts as a negative regulator under normal conditions but initiates hypertrophic signaling in the myocardium when it is oxidized due to stress [66, 67]. Moreover, the important role of Trx in hypertension has been demonstrated in mice, through experiments involving injection of human Trx and other studies in which Trx-overexpressing transgenic mice were assessed [68]. In these studies, Trx decreased age-related hypertension through different mechanisms, such as preservation of functional eNOS and NO release, and maintenance of relaxation responses. These researchers also found decreased arterial stiffness and improved vascular flow as a result of both Trx injection and overexpression. In another study, Trx1 overexpression was found to decrease infarct size and improve myocardial function after infarction [69].

The Grx superfamily has also been associated with enhanced cell survival and improved resistance to OS in CVD. In fact, overexpression of Grx1 has been shown to reduce ventricular remodeling and improve cardiac function [70], effects that were later proposed to be associated with the induction of angiogenesis. Nevertheless, Grx may offer protection to the ischemic heart by inhibiting apoptosis as demonstrated by an *in vivo* study in which Grx up-regulation inhibited EC migration and impaired hind limb revascularization [71]. The cardioprotective effect of Grx1 via a decrease in OS-mediated apoptosis is well documented [72–74]. In human coronary arteries, Grx was expressed (together with Trx) in ECs, fibroblasts, and smooth muscle cells. Specifically, the macrophages infiltrating fibrous plaques of atheroma that are known to produce ROS strongly express Grx and Trx [75]. In the year 2000, a Grx isoform located in the cytosol was discovered, defined as Grx3 or protein kinase C-interacting cousin of thioredoxin (PICOT). Although there is little data available about this oxidoreductase, it has a promising cardioprotective effect against ischemia/reperfusion-induced cardiac injury [76], as well as in aging and heart failure [77].

Apart from oxidoreductases, Prx proteins are essential for the regulation of OS and H2O2-mediated intracellular signaling. Some studies have shown that different isoforms of Prx protect against atherosclerosis. On the one hand, Prx1 scavenges H2O2 [78] and degrades lipids in macrophages, thereby reducing foam cell formation [79]. On the other hand, Prx2 and Prx4 modify the development of atherosclerotic plaques by altering immune cell infiltration [80, 81]. Moreover, Prx2 protects against neointimal thickening [82] and negatively regulates H2O2 generation and the formation of thrombosis [83, 84], whereas, its deficiency has been shown to promote the progression of the abdominal aortic aneurysm [85]. Indeed, Prx has an important influence on myocardial tissue and the overexpression of Prdx1 in cardiomyocytes was recently proposed to confer protection against cardiac hypertrophy and heart failure in the presence of pressure overload [86]. Prx3 ameliorates inflammation and protects the heart against left ventricular remodeling and failure after myocardial infarction [87], and both Prx3 and Prx6 may have a protective role during ischemia/reperfusion-induced heart injury [88]. Together these data suggest that Prx could be a potential target in therapeutic strategies to combat CVD.

4. Redox proteomics to get inside oxidative stress

The regulation of redox signaling commonly involves the post-translational modification (PTM) of proteins, in which different amino acids, including methionine and tyrosine, and most importantly Cys, are sensitive to oxidation and other modifications [89, 90]. As described above, the oxidation/reduction of thiol proteins has emerged as one of the major mechanisms through which reactive oxidants modulate cell signaling [91]. Accordingly, several proteomics-based techniques have been developed to enrich, identify, and characterize thiol-related redox modifications, referred to as redox proteomics [92]. The main aims of redox proteomics include (1) the identification of proteins that are modified or that interact with regulatory oxidoreductases, (2) the identification of residues susceptible to modification, including specific ROS-induced redox modifications, and (3) the quantification of the proportions of the modified protein(s) [93]. Crucially, RS^- is very reactive, and thus in order to study reversible Cys oxidations, it is essential to block this reactivity to capture the *in vivo* thiol-redox status. For this reason, one of the initial steps in thiol proteomics is to block free thiols in a sample with an alkylating reagent, such as iodoacetamide or derivatives of maleimide [94]. This also enables the reversibility of thiol modifications to be used to selectively label and/or purify redox-modified proteins. After alkylation, proteins can then be reduced using a thiol reducing agent, such as dithiothreitol or tris (2-carboxyethyl) phosphine, and, newly-formed free thiols can also be studied. Differential thiol isotopic labeling allows samples of different groups to be combined, reducing run-to-run experimental variation and permitting relative quantification between different samples (**Figure 3**) [95]. Although redox proteomics is still in its infancy, assessment of possible biomarkers through mass spectrometry (MS)-driven strategies may be promising for the study of OS in different disease states. In contrast to direct measurement of ROS levels, which is a complex task given the short half-life and high reactivity of these species, redox proteomics allows the detection of the resulting oxidative damage to proteins. Thus, it provides mechanistic insight into cellular toxicity mechanisms. For that reason, some studies employing this approach have been performed in CVDs.

One of the first attempts to apply redox proteome analysis to heart proteins was performed using two different proteomic methods: gel-based (DIGE) and gel-free proteomics (ICAT) [96]. Heart tissue was analyzed with the purpose to identify and quantify redox-sensitive proteins, resulting in the elucidation of 50 proteins as potential targets of H2O2 oxidation through the ICAT method and 26 such proteins with the DIGE method, 13 of which were detected with both methods. Several years later, a different technique, named GELSILOX, was used to investigate the mechanisms of damage produced in the heart by ischemia/reperfusion injury and the effects of ischemic preconditioning in mitochondria purified from cardiomyocytes [97]. Several proteins were identified that had previously been associated with redox changes but had not been distinguished in the context of ischemia/reperfusion damage. Although these were mainly proof-of-concept studies of the methodology used, they provide useful information

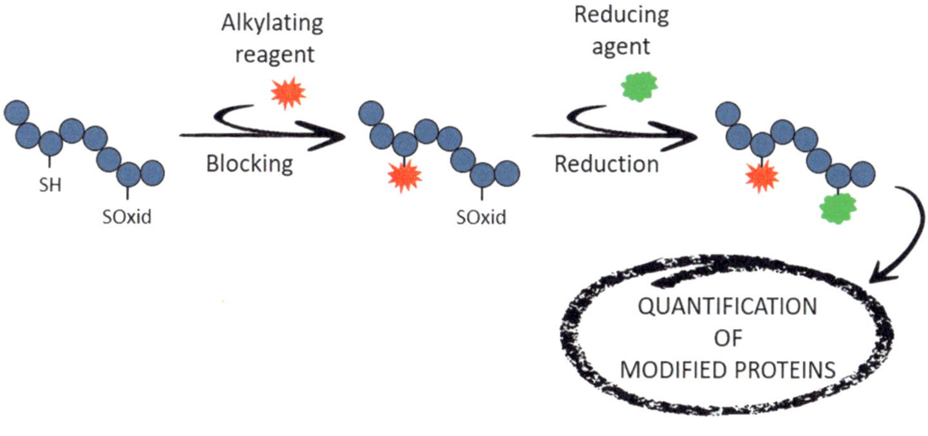

Figure 3.
Essential workflow for redox proteomics analysis includes blocking or alkylation of free thiols and reduction to form new free thiols that can also be studied. This allows the identification of proteins that are modified and the quantification of the relevant proportions of the modified protein.

about the redox proteome of the cardiovascular system, including specific sites of oxidation. More recently, the plasma redox proteome was analyzed using FASILOX, a novel multiplexed proteomic strategy of isotope labeling, followed by liquid chromatography (LC)-tandem MS [98]. A plasma signature was proposed for the stratification of young individuals and to detect those with a higher probability of suffering a future cardiovascular event. The same approach was used to analyze aortic valve tissue obtained from patients with calcific aortic valve disease [99], which revealed different protein profiles in calcified valve tissue in patients with and without atherosclerosis. Differences in the redox status of the aortic valve tissue were also found despite their high degree of calcification and affectation. Additionally, two specific sites of cysteine oxidation in albumin that had not been described previously were also found. The results of these studies highlight the enormous potential of redox proteomics, an approach that is set to become a key tool to obtain new insights into CVD-related protein modifications.

5. Future perspectives: a focus on redox biology to improve patient management

Overall, excessive OS due to exaggerated ROS production or insufficient antioxidant defense has been closely associated with CVD. Improving our understanding of these pathways may permit the modulation of redox biology, which can be instrumental in reducing the systemic impact of these alterations. From a clinical point of view, the combination of standard clinical evaluation with results from high throughput approaches is essential to take a step forward in precision medicine, currently a highly desired goal in medicine. By definition, precision medicine focuses on selecting the appropriate treatment for each patient based on individual phenotypes. This includes a reliable and accurate risk stratification that would allow better screening of patients at high risk and avoiding the use of unnecessary treatments with regard to potential side effects. Since the disruption in OS balance leads to a vicious cycle, it is crucial to identify high-risk patients to avoid the damage caused by an uncontrolled redox milieu.

CVDs are multifactorial diseases that are usually accompanied by other comorbidities, and OS incites a systemic effect that can damage different organs and systems. From a pharmacological point of view, this implies that modification of the redox state may also affect distinct organs in different patients, which may have beneficial and detrimental effects. This kind of treatment can theoretically improve the function of different systems and may be considered an integral treatment, although this unfortunately leads to less control over potential side effects. The only way to overcome this issue is to intensify research focusing on OS and its relationship with disease. Although molecules related to OS are well known but their possible roles as cardioprotective molecules are still not fully understood. More information about this facet of their activity will be essential to design successful new pharmacological approaches. However, we must be aware that there is a long way to go in this regard. Indeed, it should not be forgotten that alterations that alter multiple metabolic pathways may be detrimental when applied at inappropriate times or doses.

Acknowledgements

This research was funded by the Junta de Comunidades de Castilla-La Mancha (JCCM, co-funded by the European Regional Development Fund,

SBPLY/19/180501/000226), the Instituto de Salud Carlos III through the project PI18/00995, PI21/00384 (co-funded by European Regional Development Fund/European Social Fund—"Investing in your future") Sociedad Española de Cardiología, 2020 and Grant PRB3 (IPT17/0019—ISCIII-SGEFI/ERDF). These results are aligned with the Spanish initiative on the Human Proteome Project (SpHPP).

Conflict of interest

The authors declare no conflict of interest.

Author details

Laura Mourino-Alvarez, Tamara Sastre-Oliva, Nerea Corbacho-Alonso and Maria G. Barderas*
Department of Vascular Physiopathology, Hospital Nacional de Paraplejicos, SESCAM, Toledo, Spain

*Address all correspondence to: megonzalezb@sescam.jccm.es; marucitos@hotmail.com

References

[1] Collaborators GDaI. Global burden of 369 diseases and injuries in 204 countries and territories, 1990-2019: A systematic analysis for the global burden of disease study 2019. Lancet. 2020;**396**(10258):1204-1222. DOI: 10.1016/s0140-6736(20)30925-9

[2] Mensah GA, Roth GA, Fuster V. The global burden of cardiovascular diseases and risk factors: 2020 and beyond. Journal of the American College of Cardiology. 2019;**74**(20):2529-2532. DOI: 10.1016/j.jacc.2019.10.009

[3] Roth GA, Mensah GA, Johnson CO, Addolorato G, Ammirati E, Baddour LM, et al. Global burden of cardiovascular diseases and risk factors, 1990-2019: Update from the GBD 2019 study. Journal of the American College of Cardiology. 2020;**76**(25):2982-3021. DOI: 10.1016/j.jacc.2020.11.010

[4] Brunelli E, La Russa D, Pellegrino D. Impaired oxidative status is strongly associated with cardiovascular risk factors. Oxidative medicine and cellular longevity. 2017;**2017**:6480145. DOI: 10.1155/2017/6480145

[5] Fuster V, Voûte J. MDGs: Chronic diseases are not on the agenda. Lancet. 2005;**366**(9496):1512-1514. DOI: 10.1016/s0140-6736(05)67610-6

[6] Lepedda AJ, Formato M. Oxidative modifications in advanced atherosclerotic plaques: A focus on in situ protein sulfhydryl group oxidation. Oxidative medicine and cellular longevity. 2020;**2020**:6169825. DOI: 10.1155/2020/6169825

[7] Pastore A, Piemonte F. Protein glutathionylation in cardiovascular diseases. International Journal of Molecular Sciences. 2013;**14**(10):20845-20876. DOI: 10.3390/ijms141020845

[8] Halliwell B. The role of oxygen radicals in human disease, with particular reference to the vascular system. Haemostasis. 1993;**23**(Suppl. 1): 118-126. DOI: 10.1159/000216921

[9] Zhao S, Cheng CK, Zhang CL, Huang Y. Interplay between oxidative stress, cyclooxygenases, and prostanoids in cardiovascular diseases. Antioxidants & Redox Signaling. 2021;**34**(10):784-799. DOI: 10.1089/ars.2020.8105

[10] Xu T, Ding W, Ji X, Ao X, Liu Y, Yu W, et al. Oxidative stress in cell death and cardiovascular diseases. Oxidative Medicine and Cellular Longevity. 2019;**2019**:9030563. DOI: 10.1155/2019/9030563

[11] Sena CM, Pereira AM, Seiça R. Endothelial dysfunction—A major mediator of diabetic vascular disease. Biochimica et Biophysica Acta. 2013;**2013**(1832):2216-2231. DOI: 10.1016/j.bbadis.2013.08.006

[12] Case AJ, Tian J, Zimmerman MC. Increased mitochondrial superoxide in the brain, but not periphery, sensitizes mice to angiotensin II-mediated hypertension. Redox Biology. 2017;**11**:82-90. DOI: 10.1016/j.redox.2016.11.011

[13] Lopes RA, Neves KB, Tostes RC, Montezano AC, Touyz RM. Downregulation of nuclear factor erythroid 2-related factor and associated antioxidant genes contributes to redox-sensitive vascular dysfunction in hypertension. Hypertension. 2015;**66**(6):1240-1250. DOI: 10.1161/hypertensionaha.115.06163

[14] Silva GC, Silva JF, Diniz TF, Lemos VS, Cortes SF. Endothelial dysfunction in DOCA-salt-hypertensive mice: Role of neuronal nitric oxide synthase-derived hydrogen peroxide. Clinical Science (London, England). 2016;**130**(11):895-906. DOI: 10.1042/cs20160062

[15] Guzik TJ, Touyz RM. Oxidative stress, inflammation, and vascular aging in hypertension. Hypertension. 2017;**70**(4):660-667. DOI: 10.1161/hypertensionaha.117.07802

[16] Cai H, Harrison DG. Endothelial dysfunction in cardiovascular diseases: The role of oxidant stress. Circulation Research. 2000;**87**(10):840-844. DOI: 10.1161/01.res.87.10.840

[17] Brown DI, Griendling KK. Regulation of signal transduction by reactive oxygen species in the cardiovascular system. Circulation Research. 2015;**116**(3):531-549. DOI: 10.1161/circresaha.116.303584

[18] Gao L, Siu KL, Chalupsky K, Nguyen A, Chen P, Weintraub NL, et al. Role of uncoupled endothelial nitric oxide synthase in abdominal aortic aneurysm formation: Treatment with folic acid. Hypertension. 2012;**59**(1):158-166. DOI: 10.1161/hypertensionaha.111.181644

[19] Youn JY, Wang T, Blair J, Laude KM, Oak JH, McCann LA, et al. Endothelium-specific sepiapterin reductase deficiency in DOCA-salt hypertension. American Journal of Physiology. Heart and Circulatory Physiology. 2012;**302**(11):H2243-H2249. DOI: 10.1152/ajpheart.00835.2011

[20] Siu KL, Miao XN, Cai H. Recoupling of eNOS with folic acid prevents abdominal aortic aneurysm formation in angiotensin II-infused apolipoprotein E null mice. PLoS One. 2014;**9**(2):e88899. DOI: 10.1371/journal.pone.0088899

[21] Förstermann U, Closs EI, Pollock JS, Nakane M, Schwarz P, Gath I, et al. Nitric oxide synthase isozymes. Characterization, purification, molecular cloning, and functions. Hypertension. 1994;**23**(6 Pt 2):1121-1131. DOI: 10.1161/01.hyp.23.6.1121

[22] Daiber A, Xia N, Steven S, Oelze M, Hanf A, Kröller-Schön S, et al. New therapeutic implications of endothelial nitric oxide synthase (eNOS) function/dysfunction in cardiovascular disease. International Journal of Molecular Sciences. 2019;**20**(1):187. DOI: 10.3390/ijms200101877

[23] Zorov DB, Filburn CR, Klotz LO, Zweier JL, Sollott SJ. Reactive oxygen species (ROS)-induced ROS release: A new phenomenon accompanying induction of the mitochondrial permeability transition in cardiac myocytes. The Journal of Experimental Medicine. 2000;**192**(7):1001-1014. DOI: 10.1084/jem.192.7.1001

[24] Zinkevich NS, Gutterman DD. ROS-induced ROS release in vascular biology: Redox-redox signaling. American Journal of Physiology. Heart and Circulatory Physiology. 2011;**301**(3):H647-H653. DOI: 10.1152/ajpheart.01271.2010

[25] Kim YM, Kim SJ, Tatsunami R, Yamamura H, Fukai T, Ushio-Fukai M. ROS-induced ROS release orchestrated by Nox4, Nox2, and mitochondria in VEGF signaling and angiogenesis. American Journal of Physiology Cell Physiology. 2017;**312**(6):C749-Cc64. DOI: 10.1152/ajpcell.00346.2016

[26] Murphy MP. How mitochondria produce reactive oxygen species. The Biochemical Journal. 2009;**417**(1):1-13. DOI: 10.1042/bj20081386

[27] Andreyev AY, Kushnareva YE, Starkov AA. Mitochondrial metabolism of reactive oxygen species. Biochemistry Biokhimiia. 2005;**70**(2):200-214. DOI: 10.1007/s10541-005-0102-7

[28] Starkov AA. The role of mitochondria in reactive oxygen species metabolism and signaling. Annals of the New York Academy of Sciences. 2008;**1147**:37-52. DOI: 10.1196/annals.1427.015

[29] Förstermann U, Xia N, Li H. Roles of vascular oxidative stress and nitric oxide in the pathogenesis of atherosclerosis. Circulation Research. 2017;**120**(4):713-735. DOI: 10.1161/circresaha.116.309326

[30] Tward A, Xia YR, Wang XP, Shi YS, Park C, Castellani LW, et al. Decreased atherosclerotic lesion formation in human serum paraoxonase transgenic mice. Circulation. 2002;**106**(4):484-490. DOI: 10.1161/01.cir.0000023623.87083.4f

[31] Yang H, Roberts LJ, Shi MJ, Zhou LC, Ballard BR, Richardson A, et al. Retardation of atherosclerosis by overexpression of catalase or both Cu/Zn-superoxide dismutase and catalase in mice lacking apolipoprotein E. Circulation Research. 2004;**95**(11):1075-1081. DOI: 10.1161/01.RES.0000149564.49410.0d

[32] Dikalova AE, Itani HA, Nazarewicz RR, McMaster WG, Flynn CR, Uzhachenko R, et al. Sirt3 impairment and SOD2 hyperacetylation in vascular oxidative stress and hypertension. Circulation Research. 2017;**121**(5):564-574. DOI: 10.1161/circresaha.117.310933

[33] Su H, Zeng H, Liu B, Chen JX. Sirtuin 3 is essential for hypertension-induced cardiac fibrosis via mediating pericyte transition. Journal of Cellular and Molecular Medicine. 2020;**24**(14):8057-8068. DOI: 10.1111/jcmm.15437

[34] Mercer JR, Yu E, Figg N, Cheng KK, Prime TA, Griffin JL, et al. The mitochondria-targeted antioxidant MitoQ decreases features of the metabolic syndrome in ATM+/−/ApoE−/− mice. Free Radical Biology & Medicine. 2012;**52**(5):841-849. DOI: 10.1016/j.freeradbiomed.2011.11.026

[35] Graham D, Huynh NN, Hamilton CA, Beattie E, Smith RA, Cochemé HM, et al. Mitochondria-targeted antioxidant MitoQ10 improves endothelial function and attenuates cardiac hypertrophy. Hypertension. 2009;**54**(2):322-328. DOI: 10.1161/hypertensionaha.109.130351

[36] Rossman MJ, Santos-Parker JR, Steward CAC, Bispham NZ, Cuevas LM, Rosenberg HL, et al. Chronic supplementation with a mitochondrial antioxidant (MitoQ) improves vascular function in healthy older adults. Hypertension. 2018;**71**(6):1056-1063. DOI: 10.1161/hypertensionaha.117.10787

[37] Young CN. Endoplasmic reticulum stress in the pathogenesis of hypertension. Experimental Physiology. 2017;**102**(8):869-884. DOI: 10.1113/ep086274

[38] Zito E. ERO1: A protein disulfide oxidase and H2O2 producer. Free Radical Biology & Medicine. 2015;**83**:299-304. DOI: 10.1016/j.freeradbiomed.2015.01.011

[39] Rainbolt TK, Saunders JM, Wiseman RL. Stress-responsive regulation of mitochondria through the ER unfolded protein response. Trends in Endocrinology and Metabolism: TEM. 2014;**25**(10):528-537. DOI: 10.1016/j.tem.2014.06.007

[40] Malhotra JD, Kaufman RJ. Endoplasmic reticulum stress and oxidative stress: A vicious cycle or a double-edged sword? Antioxidants & Redox Signaling. 2007;**9**(12):2277-2293. DOI: 10.1089/ars.2007.1782

[41] Cao SS, Kaufman RJ. Endoplasmic reticulum stress and oxidative stress in cell fate decision and human disease. Antioxidants & Redox Signaling. 2014;**21**(3):396-413. DOI: 10.1089/ars.2014.5851

[42] Bravo R, Gutierrez T, Paredes F, Gatica D, Rodriguez AE, Pedrozo Z, et al. Endoplasmic reticulum: ER stress regulates mitochondrial bioenergetics. The International Journal of Biochemistry & Cell Biology. 2012;**44**(1):16-20. DOI: 10.1016/j.biocel.2011.10.012

[43] Santos CX, Nabeebaccus AA, Shah AM, Camargo LL, Filho SV, Lopes LR. Endoplasmic reticulum stress and Nox-mediated reactive oxygen species signaling in the peripheral vasculature: Potential role in hypertension. Antioxidants & Redox Signaling. 2014;**20**(1):121-134. DOI: 10.1089/ars.2013.5262

[44] Shanahan CM, Furmanik M. Endoplasmic reticulum stress in arterial smooth muscle cells: A novel regulator of vascular disease. Current Cardiology Reviews. 2017;**13**(2):94-105. DOI: 10.2174/1573403x12666161014094738

[45] Ochoa CD, Wu RF, Terada LS. ROS signaling and ER stress in cardiovascular disease. Molecular Aspects of Medicine. 2018;**63**:18-29. DOI: 10.1016/j.mam.2018.03.002

[46] Neuber C, Uebeler J, Schulze T, Sotoud H, El-Armouche A, Eschenhagen T. Guanabenz interferes with ER stress and exerts protective effects in cardiac myocytes. PLoS One. 2014;**9**(6):e98893. DOI: 10.1371/journal.pone.0098893

[47] Abdulkarim B, Hernangomez M, Igoillo-Esteve M, Cunha DA, Marselli L, Marchetti P, et al. Guanabenz sensitizes pancreatic β cells to lipotoxic endoplasmic reticulum stress and apoptosis. Endocrinology. 2017;**158**(6):1659-1670. DOI: 10.1210/en.2016-1773

[48] Choi SK, Lim M, Byeon SH, Lee YH. Inhibition of endoplasmic reticulum stress improves coronary artery function in the spontaneously hypertensive rats. Scientific Reports. 2016;**6**:31925. DOI: 10.1038/srep31925

[49] Ceylan-Isik AF, Sreejayan N, Ren J. Endoplasmic reticulum chaperon tauroursodeoxycholic acid alleviates obesity-induced myocardial contractile dysfunction. Journal of Molecular and Cellular Cardiology. 2011;**50**(1):107-116. DOI: 10.1016/j.yjmcc.2010.10.023

[50] Han S, Bal NB, Sadi G, Usanmaz SE, Tuglu MM, Uludag MO, et al. Inhibition of endoplasmic reticulum stress protected DOCA-salt hypertension-induced vascular dysfunction. Vascular Pharmacology. 2019;**113**:38-46. DOI: 10.1016/j.vph.2018.11.004

[51] Lynn EG, Lhoták Š, Lebeau P, Byun JH, Chen J, Platko K, et al. 4-Phenylbutyrate protects against atherosclerotic lesion growth by increasing the expression of HSP25 in macrophages and in the circulation of Apoe(−/−) mice. The FASEB Journal. 2019;**33**(7):8406-8422. DOI: 10.1096/fj.201802293RR

[52] Luo T, Kim JK, Chen B, Abdel-Latif A, Kitakaze M, Yan L. Attenuation of ER stress prevents post-infarction-induced cardiac rupture

and remodeling by modulating both cardiac apoptosis and fibrosis. Chemico-Biological Interactions. 2015;**225**:90-98. DOI: 10.1016/j.cbi.2014.10.032

[53] Luo T, Chen B, Wang X. 4-PBA prevents pressure overload-induced myocardial hypertrophy and interstitial fibrosis by attenuating endoplasmic reticulum stress. Chemico-Biological Interactions. 2015;**242**:99-106. DOI: 10.1016/j.cbi.2015.09.025

[54] Holmgren A. Antioxidant function of thioredoxin and glutaredoxin systems. Antioxidants & Redox Signaling. 2000;**2**(4):811-820. DOI: 10.1089/ars.2000.2.4-811

[55] Berndt C, Lillig CH, Holmgren A. Thiol-based mechanisms of the thioredoxin and glutaredoxin systems: Implications for diseases in the cardiovascular system. American Journal of Physiology. Heart and Circulatory Physiology. 2007;**292**(3):H1227-H1236. DOI: 10.1152/ajpheart.01162.2006

[56] Rhee SG, Kang SW, Chang TS, Jeong W, Kim K. Peroxiredoxin, a novel family of peroxidases. IUBMB Life. 2001;**52**(1-2):35-41. DOI: 10.1080/15216540252774748

[57] Szeliga M. Peroxiredoxins in neurodegenerative diseases. Antioxidants (Basel, Switzerland). 2020;**9**(12):1203. DOI: 10.3390/antiox9121203

[58] Jeong SJ, Park JG, Oh GT. Peroxiredoxins as potential targets for cardiovascular disease. Antioxidants (Basel, Switzerland). 2021;**10**(8):1244. DOI: 10.3390/antiox10081244

[59] Andreadou I, Efentakis P, Frenis K, Daiber A, Schulz R. Thiol-based redox-active proteins as cardioprotective therapeutic agents in cardiovascular diseases. Basic Research in Cardiology. 2021;**116**(1):44. DOI: 10.1007/s00395-021-00885-5

[60] Aota M, Matsuda K, Isowa N, Wada H, Yodoi J, Ban T. Protection against reperfusion-induced arrhythmias by human thioredoxin. Journal of Cardiovascular Pharmacology. 1996;**27**(5):727-732. DOI: 10.1097/00005344-199605000-00016

[61] Yoshioka J, Schulze PC, Cupesi M, Sylvan JD, MacGillivray C, Gannon J, et al. Thioredoxin-interacting protein controls cardiac hypertrophy through regulation of thioredoxin activity. Circulation. 2004;**109**(21):2581-2586. DOI: 10.1161/01.cir.0000129771.32215.44

[62] Meuillet EJ, Mahadevan D, Berggren M, Coon A, Powis G. Thioredoxin-1 binds to the C2 domain of PTEN inhibiting PTEN's lipid phosphatase activity and membrane binding: A mechanism for the functional loss of PTEN's tumor suppressor activity. Archives of Biochemistry and Biophysics. 2004;**429**(2):123-133. DOI: 10.1016/j.abb.2004.04.020

[63] Turoczi T, Chang VW, Engelman RM, Maulik N, Ho YS, Das DK. Thioredoxin redox signaling in the ischemic heart: An insight with transgenic mice overexpressing Trx1. Journal of Molecular and Cellular Cardiology. 2003;**35**(6):695-704. DOI: 10.1016/s0022-2828(03)00117-2

[64] Jekell A, Hossain A, Alehagen U, Dahlström U, Rosén A. Elevated circulating levels of thioredoxin and stress in chronic heart failure. European Journal of Heart Failure. 2004;**6**(7):883-890. DOI: 10.1016/j.ejheart.2004.03.003

[65] Kishimoto C, Shioji K, Nakamura H, Nakayama Y, Yodoi J, Sasayama S. Serum thioredoxin (TRX) levels in patients with heart failure. Japanese Circulation

Journal. 2001;**65**(6):491-494. DOI: 10.1253/jcj.65.491

[66] Ago T, Sadoshima J. Thioredoxin1 as a negative regulator of cardiac hypertrophy. Antioxidants & Redox Signaling. 2007;**9**(6):679-687. DOI: 10.1089/ars.2007.1529

[67] Haendeler J, Popp R, Goy C, Tischler V, Zeiher AM, Dimmeler S. Cathepsin D and H2O2 stimulate degradation of thioredoxin-1: Implication for endothelial cell apoptosis. The Journal of Biological Chemistry. 2005;**280**(52):42945-42951. DOI: 10.1074/jbc.M506985200

[68] Hilgers RH, Kundumani-Sridharan V, Subramani J, Chen LC, Cuello LG, Rusch NJ, et al. Thioredoxin reverses age-related hypertension by chronically improving vascular redox and restoring eNOS function. Science Translational Medicine. 2017;**9**(376):eaaf6094. DOI: 10.1126/scitranslmed.aaf6094

[69] Koneru S, Penumathsa SV, Thirunavukkarasu M, Zhan L, Maulik N. Thioredoxin-1 gene delivery induces heme oxygenase-1 mediated myocardial preservation after chronic infarction in hypertensive rats. American Journal of Hypertension. 2009;**22**(2):183-190. DOI: 10.1038/ajh.2008.318

[70] Adluri RS, Thirunavukkarasu M, Zhan L, Dunna NR, Akita Y, Selvaraju V, et al. Glutaredoxin-1 overexpression enhances neovascularization and diminishes ventricular remodeling in chronic myocardial infarction. PLoS One. 2012;**7**(3):e34790. DOI: 10.1371/journal.pone.0034790

[71] Murdoch CE, Shuler M, Haeussler DJ, Kikuchi R, Bearelly P, Han J, et al. Glutaredoxin-1 up-regulation induces soluble vascular endothelial growth factor receptor 1, attenuating post-ischemia limb revascularization. The Journal of Biological Chemistry. 2014;**289**(12):8633-8644. DOI: 10.1074/jbc.M113.517219

[72] Gallogly MM, Shelton MD, Qanungo S, Pai HV, Starke DW, Hoppel CL, et al. Glutaredoxin regulates apoptosis in cardiomyocytes via NFkappaB targets Bcl-2 and Bcl-xL: Implications for cardiac aging. Antioxidants & Redox Signaling. 2010;**12**(12):1339-1353. DOI: 10.1089/ars.2009.2791

[73] Murata H, Ihara Y, Nakamura H, Yodoi J, Sumikawa K, Kondo T. Glutaredoxin exerts an antiapoptotic effect by regulating the redox state of Akt. The Journal of Biological Chemistry. 2003;**278**(50):50226-50233. DOI: 10.1074/jbc.M310171200

[74] Pan S, Berk BC. Glutathiolation regulates tumor necrosis factor-alpha-induced caspase-3 cleavage and apoptosis: Key role for glutaredoxin in the death pathway. Circulation Research. 2007;**100**(2):213-219. DOI: 10.1161/01.res.0000256089.30318.20

[75] Okuda M, Inoue N, Azumi H, Seno T, Sumi Y, Hirata K, et al. Expression of glutaredoxin in human coronary arteries: Its potential role in antioxidant protection against atherosclerosis. Arteriosclerosis, Thrombosis, and Vascular Biology. 2001;**21**(9):1483-1487. DOI: 10.1161/hq0901.095550

[76] Kim J, Kim J, Kook H, Park WJ. PICOT alleviates myocardial ischemia-reperfusion injury by reducing intracellular levels of reactive oxygen species. Biochemical and Biophysical Research Communications. 2017;**485**(4):807-813. DOI: 10.1016/j.bbrc.2017.02.136

[77] Donelson J, Wang Q, Monroe TO, Jiang X, Zhou J, Yu H, et al.

Cardiac-specific ablation of glutaredoxin 3 leads to cardiac hypertrophy and heart failure. Physiological Reports. 2019;7(8):e14071. DOI: 10.14814/phy2.14071

[78] Kisucka J, Chauhan AK, Patten IS, Yesilaltay A, Neumann C, Van Etten RA, et al. Peroxiredoxin1 prevents excessive endothelial activation and early atherosclerosis. Circulation Research. 2008;**103**(6):598-605. DOI: 10.1161/circresaha.108.174870

[79] Jeong SJ, Kim S, Park JG, Jung IH, Lee MN, Jeon S, et al. Prdx1 (peroxiredoxin 1) deficiency reduces cholesterol efflux via impaired macrophage lipophagic flux. Autophagy. 2018;**14**(1):120-133. DOI: 10.1080/15548627.2017.1327942

[80] Guo X, Yamada S, Tanimoto A, Ding Y, Wang KY, Shimajiri S, et al. Overexpression of peroxiredoxin 4 attenuates atherosclerosis in apolipoprotein E knockout mice. Antioxidants & Redox Signaling. 2012;**17**(10):1362-1375. DOI: 10.1089/ars.2012.4549

[81] Park JG, Yoo JY, Jeong SJ, Choi JH, Lee MR, Lee MN, et al. Peroxiredoxin 2 deficiency exacerbates atherosclerosis in apolipoprotein E-deficient mice. Circulation Research. 2011;**109**(7):739-749. DOI: 10.1161/circresaha.111.245530

[82] Choi MH, Lee IK, Kim GW, Kim BU, Han YH, Yu DY, et al. Regulation of PDGF signalling and vascular remodelling by peroxiredoxin II. Nature. 2005;**435**(7040):347-353. DOI: 10.1038/nature03587

[83] Jang JY, Wang SB, Min JH, Chae YH, Baek JY, Yu DY, et al. Peroxiredoxin II is an antioxidant enzyme that negatively regulates collagen-stimulated platelet function. The Journal of Biological Chemistry. 2015;**290**(18):11432-11442. DOI: 10.1074/jbc.M115.644260

[84] Li W, Febbraio M, Reddy SP, Yu DY, Yamamoto M, Silverstein RL. CD36 participates in a signaling pathway that regulates ROS formation in murine VSMCs. The Journal of Clinical Investigation. 2010;**120**(11):3996-4006. DOI: 10.1172/jci42823

[85] Jeong SJ, Cho MJ, Ko NY, Kim S, Jung IH, Min JK, et al. Deficiency of peroxiredoxin 2 exacerbates angiotensin II-induced abdominal aortic aneurysm. Experimental & Molecular Medicine. 2020;**52**(9):1587-1601. DOI: 10.1038/s12276-020-00498-3

[86] Tang C, Yin G, Huang C, Wang H, Gao J, Luo J, et al. Peroxiredoxin-1 ameliorates pressure overload-induced cardiac hypertrophy and fibrosis. Biomedicine & Pharmacotherapy = Biomedecine & Pharmacotherapie. 2020;**129**:110357. DOI: 10.1016/j.biopha.2020.110357

[87] Matsushima S, Ide T, Yamato M, Matsusaka H, Hattori F, Ikeuchi M, et al. Overexpression of mitochondrial peroxiredoxin-3 prevents left ventricular remodeling and failure after myocardial infarction in mice. Circulation. 2006;**113**(14):1779-1786. DOI: 10.1161/circulationaha.105.582239

[88] Kumar V, Kitaeff N, Hampton MB, Cannell MB, Winterbourn CC. Reversible oxidation of mitochondrial peroxiredoxin 3 in mouse heart subjected to ischemia and reperfusion. FEBS Letters. 2009;**583**(6):997-1000. DOI: 10.1016/j.febslet.2009.02.018

[89] Mannaa A, Hanisch FG. Redox proteomes in human physiology and disease mechanisms. Journal of Proteome Research. 2020;**19**(1):1-17. DOI: 10.1021/acs.jproteome.9b00586

[90] Reczek CR, Chandel NS. ROS-dependent signal transduction. Current Opinion in Cell Biology. 2015;**33**:8-13. DOI: 10.1016/j.ceb.2014.09.010

[91] Winterbourn CC, Hampton MB. Thiol chemistry and specificity in redox signaling. Free Radical Biology & Medicine. 2008;**45**(5):549-561. DOI: 10.1016/j.freeradbiomed.2008.05.004

[92] Wojdyla K, Rogowska-Wrzesinska A. Differential alkylation-based redox proteomics—Lessons learnt. Redox Biology. 2015;**6**:240-252. DOI: 10.1016/j.redox.2015.08.005

[93] Sheehan D, McDonagh B. The clinical potential of thiol redox proteomics. Expert Review of Proteomics. 2020;**17**(1):41-48. DOI: 10.1080/14789450.2020.1704260

[94] Hansen RE, Winther JR. An introduction to methods for analyzing thiols and disulfides: Reactions, reagents, and practical considerations. Analytical Biochemistry. 2009;**394**(2):147-158. DOI: 10.1016/j.ab.2009.07.051

[95] Bonzon-Kulichenko E, Camafeita E, López JA, Gómez-Serrano M, Jorge I, Calvo E, et al. Improved integrative analysis of the thiol redox proteome using filter-aided sample preparation. Journal of Proteomics. 2020;**214**:103624. DOI: 10.1016/j.jprot.2019.103624

[96] Fu C, Hu J, Liu T, Ago T, Sadoshima J, Li H. Quantitative analysis of redox-sensitive proteome with DIGE and ICAT. Journal of Proteome Research. 2008;7(9):3789-3802. DOI: 10.1021/pr800233r

[97] Martínez-Acedo P, Nuñez E, Gómez FJS, Moreno M, Ramos E, Izquierdo-Álvarez A, et al. A novel strategy for global analysis of the dynamic thiol redox proteome. Molecular & Cellular Proteomics. 2012;**11**(9):800-813. DOI: 10.1074/mcp.M111.016469

[98] Corbacho-Alonso N, Baldán-Martín M, López JA, Rodríguez-Sánchez E, Martínez PJ, Mourino-Alvarez L, et al. Cardiovascular risk stratification based on oxidative stress for early detection of pathology. Antioxidants & Redox Signaling. 2021;**35**(8):602-617. DOI: 10.1089/ars.2020.8254

[99] Sastre-Oliva T, Corbacho-Alonso N, Albo-Escalona D, Lopez JA, Lopez-Almodovar LF, Vázquez J, et al. The influence of coronary artery disease in the development of aortic stenosis and the importance of the albumin redox state. Antioxidants (Basel, Switzerland). 2022;**11**(2):317. DOI: 10.3390/antiox11020317

Chapter 8

Importance of Oxidative Stress Mechanism in Reproductive Functions and Infertility

Moyinoluwa Comfort Onaolapo, Samuel Chibueze Nzekwe, Lateef Okeleji Olabisi, Victor Oluwaseyi Amos, Oluwatobi Hezekiah Ajayi and Ayodeji Folorunsho Ajayi

Abstract

Oxidative stress (OS) is a term used to describe the homeostatic oxidation-favoring imbalance between the formation of reactive oxygen species (ROS) or other compounds causing oxidative stress and the countering activities/levels of enzymatic or non-enzymatic antioxidants. The role of OS in reproduction cannot be underestimated in neither health nor disease. This chapter focuses on the roles of OS in spermatogenesis, steroidogenesis and male sexual activity, and also its effects in female folliculogenesis, steroidogenesis, ovulation, luteogenesis, and pregnancy. Furthermore, OS's impact on the efficacy of Artificial Reproductive Techniques (ARTs) was assessed, and the impact of antioxidants on reproductive health and sterility were discussed in both males and females. Through available evidence, it appears that oxidative state impairs reproductive processes and causes general disruptions through inflammation, DNA damage, lipid peroxidation, protein alterations and mitochondrial dysfunction. It will be of importance to identify oxidative stress biomarkers specific for each reproductive process, and it seems that more research should be focused on epigenetic characteristics together with oxidative stress in reproductive health and infertility.

Keywords: female reproduction, health, infertility, male reproduction, oxidative stress

1. Introduction

Oxidative stress connotes the damage that occurs when the activities of reactive oxidants overwhelms the in vivo capabilities of antioxidants [1]. This state is characterized by an elevated amount of reactive species such as superoxide (O_2^-), peroxyl (RO_2), hydroperoxyl (HO_2), and hydroxyl (-OH), which are products of reaction of oxygen with various unsaturated lipids. Furthermore, since antioxidants constantly respond to the overwhelming oxidative insult, the levels of enzymatic and non-enzymatic antioxidants are often found to be decreased [2]. Among these parameters, those that demonstrate measurable changes are generally called OS biomarkers.

Previous investigations have implicated OS and its mechanisms in a number of diseases, including infertility [3].

Infertility is the inability to achieve conception following 12 months of continual unprotected copulation [4]. Infertility could be a result of dysfunctions from either of the partners or both, it was reported that about 48 million married individuals and a total of 186 million persons are affected by infertility [5, 6]. Similarly, a meta-analysis of researches on the prevalence of infertility suggests that about 10% of the world population lives with infertility [7].

While it must be mentioned that physiological levels of ROS are important for male reproductive processes, it is also essential to state that unchallenged activity of ROS will be, at least in turn, damaging to the health of the spermatozoa and gamete cells [8]. In men, free radical elevation in the ejaculate is mainly sponsored by high leukocyte counts as well as immature spermatozoa, which cause low fertility owing to lipid peroxidation, apoptosis and damage to the DNA of sperm cells [8, 9]. In addition, OS in male infertility is connected with a number of environmental and epigenetic factors like obesity, smoking, poor diet, infection, and exposure to some endocrine disruptors [8]. Furthermore, conditions like varicocele, testicular cancer, idiopathic male infertility and erectile dysfunction have been strongly associated with oxidative stress [10].

In females, disproportionate levels of anti- and pro-oxidants have been associated with conditions like polycystic ovary syndrome (PCOS), endometriosis and unexplained infertility [11]. One of the major ways in which oxidative stress impairs female fertility is its harmful effects on proteins and nucleic acids [11]. Other mechanisms involved in oxidative stress-induced reproductive dysfunction in females include peroxidation of arachidonic acids, release of inflammatory mediators, and apoptosis [12].

Reports have shown OS to be a common denominator in the majority of infertility states in males and females [13]. Therefore, clarifying the role of OS in crucial reproductive processes (spermatogenesis, ovulation, and steroidogenesis) and identifying the vital modulatory roles of antioxidants in these conditions will help to elucidate the subject of infertility which is plaguing the world and can lead to valuable recommendations.

2. Reactive oxygen species

2.1 Definition of reactive oxygen species

Reactive oxygen species (ROS) is a combined terminology ascribed to oxygen radicals, such as hydroxyl (-OH), superoxide (O_2^-), hydroperoxyl (HO_2) peroxyl (RO_2), radicals, and also, a number of non-radical oxidizing agents, such as hypochlorous acid (HOCl), hydrogen peroxide (H_2O_2), and ozone (O_3), which possess at least a single unpaired electron and can be transposed into radicals with ease [14]. They are toxic by-products of aerobic metabolism [15] and they are produced continually at basal levels during metabolic activities in the body. The body's antioxidant system scavenges ROS and has the capability to neutralize potential harm under physiological conditions [16]. In addition, the redox homeostasis within the cell is maintained by the reducing nature of the internal environment of the cell, thereby preventing injury as a result of free radicals. The environment however, is sustained by a number of antioxidant substances and enzymes, which include glutathione peroxidase, superoxide dismutase (SOD), glutathione, other thiols, thioredoxin, ascorbate (vitamin C) and tocopherol (vitamin E) [17]. Although studies have shown that ROS are involved in the pathogenesis of many diseases, they are also reported to be pertinent in a number of functions including signal transduction, gene expression and mitochondrial electron transport [18].

2.2 Types and sources of reactive oxygen species

The types of ROS are not to be limited to oxygen radicals (hydroxyl and superoxide), but also include some other molecular oxygen (O_2) derivatives that are non-radical, such as hydrogen peroxide (H_2O_2) [19]. Therefore, various types of ROS have been reported: lipid peroxide (LO_2) which possess no unpaired electrons, hydroxyl (OH^-), peroxyl (ROO^-), superoxide (O_2^-), alkoxy (RO) radicals, radicals of nitric oxide (NO), nitrogen dioxide (NO_2), peroxynitrite ($ONOO^-$), ozone (O_3), and perhaps singlet oxygen [2]. Even though lipid peroxide and hydrogen peroxide are exempted from the free radical list, they serve as reservoirs for peroxyl, hydroxyl and alkoxy radicals, which are very reactive. ROS are continually produced in living cells throughout life via two major pathways, which are characterized by their endogenous and/or exogenous origin(s).

2.2.1 Endogenous sources

Endogenous formation entails the production of ROS within the living organism due to cellular activities. Several enzyme groups have been implicated in catalyzing this process. The seven isoforms of the expanding family of transmembrane NADPH oxidases (NOXs), a superoxide-generating system, is a good example of such enzymes [20, 21]. There are various endogenous sources of ROS in the cell; however, the most relevant and extensive are the mitochondria, endoplasmic reticulum, and peroxisomes. In the electron transport chain (ETC), there are two major sites in which the generation of mitochondrial superoxide radicals take place, Complex I (NADH dehydrogenase) and Complex III (ubiquinone-cytochrome *c* reductase) [2, 22]. About 1–2% of consumed oxygen molecules are converted into superoxide anions at these two sites [23, 24]. In the endoplasmic reticulum, which is a lipid and protein biosynthesizing organelle, there are two main mechanisms responsible for ROS generation [25]. The initial process is the generation of ROS as a by-product during electron transfer to molecular oxygen from protein thiol structures, which is associated with protein disulfide-isomerase (PDI) and endoplasmic reticulum oxidoreductin-1 (ERO-1) [19, 25]. The other procedure entails ROS production during protein misfolding as a result of depletion of glutathione (GSH) [26, 27], this is followed by reparation of thiols thereby permitting their interaction with ERO-1/PDI and their re-oxidation [25]. The process leads to cycles of formation and breakage of disulfide bonds, and each cycle generates more ROS as a by-product [28]. Finally, in the peroxisome, ROS production takes place in diverse metabolic pathways including fatty acid α- and β-oxidation, phospholipid biosynthesis, polyamine oxidation, amino acid catabolism, and glyoxylate metabolism. Perhaps most importantly, the oxidative phase of the pentose phosphate pathway [29] functions through the activity of a diverse set of enzymes that generate several types of ROS, such as hydrogen peroxide, hydroxyl radical, superoxide, nitric oxide radicals and peroxynitrites, as part of their physiological functions [30].

2.2.2 Exogenous sources

ROS production can be prompted by a number of external/environmental factors. These include ultraviolet light, narcotic drugs, chemicals, and pollutants (in food and air) [31, 32]. When cells are exposed to radiation, it in turn leads to the production of various radical and non-radical species from ionization of intracellular water, including aqueous electrons, H_2O_2, and OH^- [19]. Air pollutants such as cigarette smoke, motor vehicle exhaust and industrial contaminants encompassing many types of NO derivatives comprise the main sources of ROS that affect and cause organism injury, either by direct contact with the skin or inhalation. Additionally, chemicals (*e.g.* paraquat) that react to form either peroxides, ozone or superoxide, and a number of drugs, such as bleomycin and adriamycin, whose mechanism of action is mediated through the generation of ROS, are also primary sources of ROS [13, 19, 33, 34]. Most notably, it is important to emphasize the fact that food is considered the most relevant source of oxidants [19]. A large proportion of consumed food is oxidized to a high extent and contain varieties of oxidants such as peroxides, oxidized fatty acids, transition metals and aldehydes. These oxidative compounds that are taken into the intestinal tract cause great oxidative pressure on the intestinal mucosa [33].

2.3 Physiological roles of ROS

Although ROS are considered detrimental to health when excessive in the body system, they also play a series of vital roles in human physiology [35]. ROS have been shown to regulate the diameter of blood vessels, where ROS from the mitochondria (specifically superoxide and hydrogen peroxide) facilitate physiological reaction to factors including shear-stress in human coronary arteries [36, 37]. Another physiologic role is their facilitation of oxygen sensing in the body [35] which is essential to cellular health due to the fact that it permits cells to initiate adaptive responses which will in turn increase the survival probability in anticipation of limited oxygen availability. The ETC in the mitochondria also acts as an oxygen sensor by producing more ROS in response to limited oxygen supply (hypoxia) [38]. Other important roles include maintenance of genomic stability and regulation of activities of the skeletal muscle [39–41]. ROS are also critical for the immune system, where the presence of pathogens results in elevated ROS generation which further results in the release of phagocytes that serve as a first defense mechanism [42].

2.4 Antioxidants

Antioxidants are the organism's means of defense against the destructive effects of ROS production and accumulation [43]. The exposure of living cells to the harmful effects of free radicals triggers reactions that activate multiple internal defense mechanisms, which helps the body in the removal of free radicals and their derivatives [44, 45]. Antioxidants engage in three major functions: preventing, repairing, and deactivating the detrimental effects of ROS [46]. Generally, antioxidants in living cells can be classified into two primary groups based on their mode of action on the ROS, enzymatic and non-enzymatic antioxidants [45].

2.4.1 Enzymatic antioxidants

Enzymatic antioxidants are antioxidants that function in the break-down and removal of free radicals. These are enzymes that convert harmful oxidative products to hydrogen peroxide (H_2O_2) and then to water in a multi-step reaction where copper, zinc, manganese, and iron are obligatory cofactors [47]. Examples of enzymatic antioxidants are superoxide dismutase (SOD), glutathione peroxidase (GPx), catalase (CAT), glutathione reductase (GR) and peroxiredoxins (Prxs) [45], which must function in concert to exert the intended antioxidant effects. SOD is particularly significant as it has the ability to catalyze the reaction that turns superoxide anion into hydrogen peroxide and molecular oxygen, which is a very relevant first line defense against ROS activity [48].

2.4.2 Non-enzymatic antioxidants

Non-enzymatic antioxidants are those that function by interrupting ROS chain reactions [47]. Few examples are vitamin E, vitamin C, carotenoids, plant polyphenol, ceruloplasmin, ferritin, thiols (e.g., glutathione) and albumin [45, 48]. Vitamin E act on cell membrane to prevent the generation of free oxygen radicals, Vitamin C prevents oxidative stress through mobbing of free oxygen radicals by neutralizing lipid hydroperoxyl radical depending of vitamin E driven mechanism and preserving proteins from alkylation through electrophilic lipid peroxidation by-products [45]. Plant polyphenols nullifying free radicals through donating of an electron or hydrogen atom [47].

3. Oxidative stress and male reproductive cells

3.1 Origin of ROS in the male reproductive system

There are different types of cells in human semen which include mature and immature sperms, leukocytes, round cells from diverse spermatogenic process stages and epithelial cells [8]. Of the aforementioned cells, the major sources of ROS are immature sperm cells and leukocytes [49]. Excessive ROS production has been reported to be associated with leukocytes (especially macrophages and neutrophils), eventually causing sperm dysfunction [8]. Reports have also shown that a positive association exists between immature sperm cells and the production of ROS. This effect may have a negative effect on ejaculate quality. Also the increase of immature sperm cells in semen is directly proportional to greater concentration of mature sperm cells with damaged DNA [50].

Apart from the aforementioned endogenous ROS sources in the reproductive system, male reproductive organs are exposed to many exogenous sources of oxidants including those derived from individual lifestyle such as alcohol use, smoking, obesity, and poor dietary intake [8]. Environmental sources of ROS include pollution, exposure to heavy metals, phthalate, heat and mobile phone radiation [8]. ROS can also affect the male reproductive system through genitourinary tract infections or could be iatrogenic through exposure to drugs, or due to clinical varicocele [8]. The role of OS in male fertility is summarized in **Figure 1**.

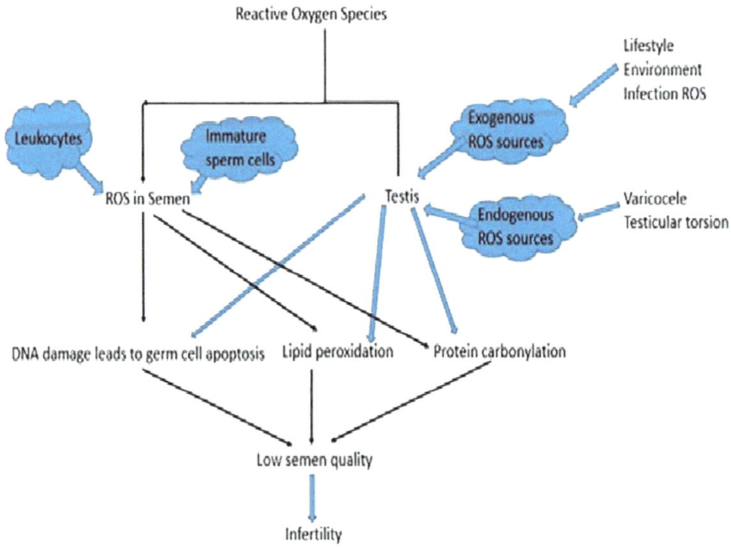

Figure 1.
Role of oxidative stress in male fertility.

3.2 Oxidative stress and spermatogenesis

Spermatogenesis involves proliferation of spermatogonia, spermatocytes meiosis and spermiogenesis occurring in the seminiferous tubules located in the testis [13]. The process is extremely replicable generating about one thousand sperm cell per second. The illustration of the process involves mitotic division of spermatogonia giving rise to spermatocytes which go through meiosis and give rise to haploid cells known as spermatids that are finally transformed by spermiation to spermatozoa [51]. When there is a disturbance in this process it can result in male infertility. One of the factors that can disrupt the process of spermatogenesis is oxidative stress [13]. Approximately, ROS contributes to about 30–80 percent of male infertility and male gametes activities are altered by oxidative stress [8].

The oxidative stress caused as a result of free radicals have a significant impact in the production as well as increasing abnormal spermatozoa, decreasing spermatozoa count and promoting sperm DNA transformation and fragmentation [52–55]. That the greater susceptibility of spermatozoa to oxidative stress when likened to other cells is owing to the fact that mature sperm cell have cytoplasm in limited amount, the sperm structure having greater level of unsaturated fatty acids and the antioxidant in sperm cells are being suppressed by ROS concentration [56]. Oxidative stress can also result in arterial occlusion then severe damage to the cell of the reproductive system and as a result defects in spermatogenesis occurs [55].

3.2.1 The antioxidant system in semen

Semen antioxidant system consist of enzymatic and non- enzymatic factors and compounds with low molecular weight having antioxidant capacity acting upon one another to bring about protection against ROS [57]. It has been reported that if any of these is inadequate it may lead to total plasma antioxidant capacity reduction [57]. Three essential antioxidant enzyme in the semen are catalase, superoxide dismutase and glutathione peroxide [57]. Superoxide dismutase (superoxide oxidoreductases – SOD) which are metaloenzymes capable of catalyzing superoxide anion dismutation reactions they are of two forms –intracellular and extracellular [58]. The intracellular forms includes copper- zinc SOD having in the active center copper and zinc (Cu, ZnSOD, SOD–1) which is found mainly in the cytoplasm, and manganese SOD found majorly in the mitochondrial matrix and having in its active center manganese (MnSOD, SOD–2). Acting in the extracellular space is the extracellular form of SOD (EC–SOD, SOD–3) and it is associated with the surface polysaccharides and can be found free, they have an active center made of copper and zinc [59].

Catalase catalyzes the reaction in which hydrogen peroxide is decomposed to molecular oxygen and water. Having a heme system structure with centrally located iron atom. It can be present in human as well as rat spermatozoa and seminal fluid having the prostrate as its source [58], and it enhances nitric oxide induced capacitation [57]. Also present as antioxidant system in the semen is the enzyme glutathione peroxidase (GPX), GPX has the capability of catalyzing organic peroxides, hydrogen peroxide as well as peroxides of phospholipids reduction [59].

3.2.2 Tests to measure reactive oxygen species in semen

Various tests which have been classified into direct and indirect assay are used in the determination of seminal ROS levels [60]. Report has shown hyperviscosity to be suggestive of oxidative stress due to its association with elevated malondialdehyde levels [60]. Factor such as increase in round cells or leukocytes being a respectable source of ROS may infer OS and abnormal sperm morphology [61]. Tests such as hypo-osmotic swelling indicate spermatozoa membrane damage as a result of lipid peroxidation therefore indicating greater ROS level in semen [62].

3.3 The effect of oxidative stress in steroidogenesis

So as to elucidate the role of oxidative stress in steroidogenesis, a number of studies have been done on animal models through introduction of exogenous sources of oxidants. ROS can impair steroidogenesis through destruction of important components in the steroidogenic pathway [63]. In another animal study, steroidogenesis, as implied by the level of FSH, LH and testosterone was seen to be reduced in animals fed with selenium-deficient diet when compared with selenium-fed animals. This indicate a possible role of oxidative stress on steroidogenesis in selenium deficient animals through the ability of selenium as an antioxidant to reverse this reduction in steriodogenesis [64]. Also increase in oxidative stress may result in gonadal dysfunction, reduction in testosterone level and testicular tissue damage indicating that gonadal steroid biosynthesis can be affected by oxidative stress [64]. In a lipopolysaccharide induced oxidative stress model in rats, it was reported that a correlation exists between a progressive oxidative state and reduction in the steroidogenic acute regulatory protein [65].

3.4 Oxidative stress and its effect on intercourse (erectile dysfunction)

Reasonably sufficient erectile and sexual functionality is essential for men [66]. Sexual activities improve the quality of life in men and promote longevity [67]. It has been reported that erectile dysfunction (ED) is a major challenge in men, which is estimated to be encountered by 40% of ≥40-year-old males [67]. ED can develop due to psychological, endocrine, vascular, neurological, and immune factors acquired via environmental exposure, lifestyle and underlying pathology [68]. The regular final mechanism of ED is vascular failure of the penis driven by significant corporal smooth muscle dysfunction which is mediated and skyrocketed via intracellular oxidative stress cumulating in a rise in smooth and endothelial muscle cell dysfunction with increase in the rate of apoptosis [67]. Nitric oxide (NO) is essential for adequate erection via nitric oxide synthase transcription (NOS) to bring about vasodilation as well as penile engorgement with blood [68]. On this basis the regular vascular failure mediator in ED is inflammation and OS, this it does through NOS reduction and subsequent reduction in NO [68]. Dysfunctional endothelial cells and increase in cellular adhesion molecules caused by inflammation enhance local arteriosclerosis and hardening of the vasculature which in turn result in local inflammation and OS [68].

3.5 Role of antioxidants in reversing oxidative stress-induced infertility in males

Genetic structure and metabolic process can influence the body's ability to produce antioxidants which has capacity to impede the effect of oxidative stress. Other factors such as pollutant, diet and chemicals can also contribute to a marked reduction in the body's antioxidant capacity [55]. Therefore,, there is need for the introduction of exogenous antioxidant to supplement the antioxidant defense in the body [55]. Therefore, the following are some of the factors reported to be free radicals scavengers and efficient antioxidants capable of reducing testicular oxidative stress [55].

3.5.1 Vitamin C and vitamin E

Vitamin E also called α-tocopherol is an effective lipophiliic antioxidant which maintains and protects spermatozoa and also contributes to the liveliness of spermatocytes and sertoli cell lines [55]. Vitamin C also called ascorbic acid plays vital role in spermatogenesis. For this reason, inadequacies in either of these two vitamins result in testicular oxidative stress and disorders of spermatogenesis and testosterone production [69]. Furthermore, vitamins C and E therapies combat oxidative stress induced by cadmium, alcohol, endosulfan and arsenic and also bring about a reduction in resultant complications [70]. Vitamin E has the ability of attenuating lipid peroxidation in mitochondrial and testicular microsomes and also able to combat adverse effects of oxidative stress that occur as a result of exposure to some exogenous factors such as iron overload, ozone gas, aflatoxin and ozone gas thereby being effectual in the protection of testicular functions [71].

3.5.2 Zinc

Zinc has been reported to be an effective antioxidant agent as well as major constituent of free radical-inhibiting enzymes like SOD [55]. Also, zinc through transferring and relocation of metals such as copper and iron is capable of preventing lipid peroxidation [72]. Studies have shown a reduction in antioxidant defense potential and a synchronous elevation in lipid peroxidation in testicular tissue in rats fed with zinc-deficient diet [70].

3.5.3 Selenium

Selenium is an important integrant of selenoproteins, it is essential in preventing OS, maintaining redox signaling state in cells and regulating thyroxine metabolism [55].

This antioxidant prevents oxidative stress by reducing free radical population in the male reproductive cells and fluids [55]. It also protect some indispensable vitamins like vitamin C and vitamin E in the body by acting synergistically with them thereby decreasing damages induced by free radicals to reproductive cells [55].

4. Effect of oxidative stress on female reproductive function

4.1 Effect of oxidative stress on folliculogenesis

It has been reported that increase in the levels ROS is related with reduction in the reproductive capability in females and infertility [73]. Reports have also shown that an elevation in the production of steroid hormone by developing follicles goes along with an increase in cytochrome P450 activity which results from the production of ROS such as hydrogen peroxide [74].

DNA damage and ovarian follicle apoptosis may be caused also by OS [75, 76]. It was noted that in dominant follicles that there is a simultaneous increase in estrogen and catalase in reaction to FSH stimulation which suggests a role of catalase in apoptosis prevention among follicles [77]. It was also reported that the oxidized form of LDL (oxLDL) and its receptor (LOX-1) are bestowed in follicular fluid or human granulosa cells and are elevated in oxidative stress states interfering with follicular maturation [78].

Growing follicles may be an inadvertent target of ROS especially in patients undergoing radiotherapy which generate a high amount of ROS [79, 80]. The ROS produced in granulosa cells may have a negative impact on oocyte fertilization as well as the rate and quality of implantation of the embryo [80]. Reports have also proven that germ cells are more vulnerable to deleterious effect of OS than somatic cells [81, 82]. Furthermore, reports has shown that OS from radiotherapy may result in ovarian atrophy, oocyte loss coupled with reduction in follicle store which may in turn result in menstrual irregularities, ovarian failure and ultimately infertility [80].

4.2 Effect of oxidative stress on steroidogenesis in females

When the ovary is over-exposed to hydrogen peroxide, it uncouples the LH receptor from adenylate cyclase. This causes disruption in protein synthesis and utilization of cholesterol by the mitochondrion p450 side chain cleavage [83]. This disruption is likely facilitated by the reduced production of steroidogenic acute regulatory protein (StAR). The StAR enhance the movement of cholesterol to the inner membrane of the mitochondria where p450 side chain cleavage converts cholesterol to pregnenolone [83]. Also, a reverse transport of cholesterol and estrogen synthesis in the follicle is facilitated by Lecithin cholesterol acyltransferase (LCAT). Evidences exist that suggest that these transporters are subjects of oxidative stress. A study reported that exogenous antioxidants like vitamin C are accumulated in mature follicles to prevent LCAT from oxidative damage and for steroidogenesis enhancement [84].

4.3 The effect of oxidative stress on ovulation

Ovulation is a process involving local inflammatory response, which leads to elevated levels of ROS [85]. The increased ROS levels may lead to potential destruction of the granulosa cells which are going through luteinization in the course of ovulation

[86]. ROS are generated during ovulation in a similar way as it occurs in inflammation. It was reported that agents that inhibits inflammatory response also suppresses ovulation [87]. The source of ROS in this process seems to be from macrophages and neutrophils because they are common in the ovaries and they led to increase production free radicals [88]. ROS has been shown to be generated during the ovulatory cascade. ROS in ovulation was noted to be mediated by protein kinase C and gonadotropin leading to production of nicotinamide adenine dinucleotide phosphate oxidase which engenders more reactive species in the course of ovulation [89].

4.4 Oxidative stress and its impact on artificial reproductive techniques (ARTs)

ARTs are advanced technological procedures which are used to treat infertility [90]. The quality of oocyte is greatly dependent on the follicular fluid microenvironment, thereby affecting the fertilization and embryo rate and quality respectively [73]. Oxidative stress markers has being said to be present in the follicular fluid of patient undertaking embryo transfer (ET) or even in-vitro fertilization (IVF) [91–94]. A reduction in intra-follicular oxygenation is said to be interrelated with a reduction in the potential of oocyte development this is as a result of an increase in the frequency of oocyte cytoplasmic disorder, impairment in cleavage and abnormal segregation in oocyte chromosome caused by follicles that are poorly vascularized [8]. The increase in embryo fragmentation, which leads to an increase in apoptosis, has been reported to be caused by ROS [8]. Hence, elevation in the ROS level is detrimental to the growth of the embryo and Sperm-oocyte interaction [8].

4.5 Role of oxidative stress in pathological pregnancies

Pathological pregnancies such as pre-eclampsia have been reported to be a complicated multisystem disorder affecting about 5–8 percent of all pregnancies and it contributes largely to fetal and maternal mortality and morbidity [95]. It has been reported that etiopathogenesis of preeclampsia may be caused by of oxidative stress and may be due to an elevation in the placenta metabolic activity as well as a reduction in its antioxidants scavenging power [95]. The role of oxidative stress in the female reproductive processes is summarized in **Figure 2**.

4.6 Role of antioxidants in reversing the effects of oxidative stress and infertility in females

Reports have shown that the female reproductive system is susceptible to oxidative damage which if left untreated the damage process continues [8, 95]. There are various antioxidants that have been reported to scavenge free radicals and to keep the reproductive system healthy [96]. They include vitamins C, E and β-carotene, L-carnitine, acetyl L-carnitine and also metallo-enzymes such as catalase, superoxide dismutase (SOD, containing copper, manganese and zinc), glutathione peroxidase (GPx, containing selenium) and superoxide dismutase (SOD, containing manganese, copper and zinc) [96].

The antioxidants mentioned above when taken helps the total antioxidant system to be coordinated functionally [97]. Antioxidant system in the ovary such as of carotenoids, CAT, glutathione and vitamin E) are responsible for regulation of ROS [96]. It was documented that the effect of SOD is noticeable in the theca interna cells of antral follicles [96]. These cells during maturation phase protect oocyte from been

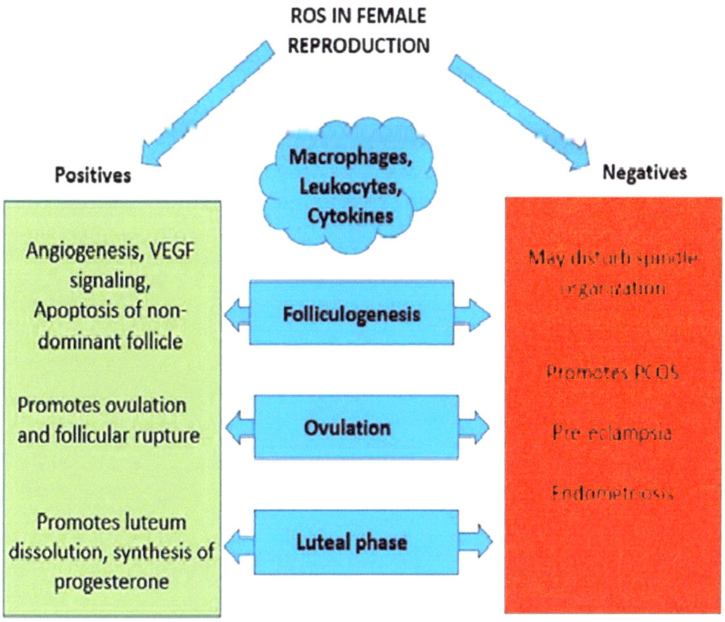

Figure 2.
Role of oxidative stress in the female reproductive processes.

destroyed by redundant ROS [96]. Vitamin C which can be found in the cytosol of oocyte and extracellular fluid is used in the treatment of luteal phase disorder and recurrent abortions [8]. Vitamin C is given to patient during in vitro fertilization (IVF) embryo transfer as a supplement in hormonal stimulation to guarantee a large concentration of vitamin C in the follicular fluid which improved oocytes and embryo qualities.

4.7 Role of oxidative stress in embryo and fetal function

During of embryonic development, the embryo is vulnerable to OS [97]. The stage of one-cell embryo depends on Krebs cycle during early phases of development, on the other hand during the other initial embryo organogenesis, anaerobic pathway and glycolysis is relied on, so does blastocyst [97]. However, there is a larger dependence on aerobic and oxidative metabolism at the establishment of circulatory system leading to a higher production of ROS by the mitochondria but antioxidants are present as well to negate and detoxify ROS [95]. But with time there may be disruption in the antioxidant and oxidant balance by the exogenous agents responsible for the stimulation of ROS which results in disruption in embryo and fetal functions [97].

5. Conclusions

Oxidative stress plays a notable role along the several processes involved in male and female reproduction. While a physiologic amount of reactive species are needed for optimal functioning of the male and female sex organs there are conditions which produces a considerable amount of reactive species and a concomitant depression of the antioxidant system. This oxidative stress state impairs the reproductive processes and causes general disruption through inflammation, DNA damage, lipid peroxidation, protein alterations and mitochondrial dysfunction.

It will be of importance to identify oxidative stress biomarkers specific for each reproductive processes and map out their standard range so as to advance measures to curtail the growing level of infertility among human population in future research.

It is also recommended that the role of genetics and oxidative stress in the etiology of infertility should be a priority for researchers.

Conflict of interest

'The Authors declare that there is no conflict of interest'.

Abbreviations

ARTs	Artificial Reproductive Techniques
CAT	Cholesterol acyltransferase
CuSOD	Copper containing superoxide dismutase
DNA	Deoxyribonucleic acid
EC–SOD or SOD–3	Extracellular form of SOD
ED	Erectile dysfunction
ERO-1	Endoplasmic reticulum oxidoreductin-1
ET	Embryo transfer
ETC	Electron transport chain
FSH	Follicle stimulating hormone
GPX	Glutathione peroxidase
GR	Glutathione reductase
GSH	Gluthatione
H_2O_2	Hydrogen Peroxide
HO_2	Hydroperoxyl
HOCl	Hypochlorous acid
IVF	In-vitro fertilization
LCAT	Lecithin cholesterol acyltransferase
LH	luteinizing hormone
LO_2	lipid peroxide
MnSOD or SOD–2	Manganese- dependent SOD
mRNA	messenger RNA
NADH	Nicotinamide adenine dinucleotide
NADPH	Nicotinamide adenine dinucleotide phosphate
NO	Nitric oxide
NO_2	Nitrogen dioxide
NOS	Nitric oxide synthase transcription (NOS)
NOXs	NADPH oxidases
O_3	Ozone
OH^-	Hydroxyl
$ONOO^-$	Peroxynitrite
OS	Oxidative stress
PCOS	Polycystic Ovary Syndrome

PDI	Protein disulfide-isomerase
Prxs	Peroxiredoxins
RO	alkoxy
ROO^-	Peroxyl
ROS	Reactive Oxygen Species
SOD	Superoxide Dismutase
SOD–1	Copper Zinc superoxide dismutase
StAR	Steroidogenic acute regulatory protein
ZnSOD	Zinc containing superoxide dismutase

Author details

Moyinoluwa Comfort Onaolapo[1], Samuel Chibueze Nzekwe[2], Lateef Okeleji Olabisi[3], Victor Oluwaseyi Amos[1], Oluwatobi Hezekiah Ajayi[1] and Ayodeji Folorunsho Ajayi[1*]

1 Department of Physiology, Ladoke Akintola University of Technology, Ogbomoso, Oyo State, Nigeria

2 Department of Biochemistry, Adeleke University, Ede, Osun State, Nigeria

3 Cardio-thoracic Unit, Obafemi Awolowo Teaching Hospital, Ile-Ife, Osun State, Nigeria

*Address all correspondence to: aajayi22@lautech.edu.ng

References

[1] Sies H. Oxidative stress: A concept in redox biology and medicine. Redox Biology. 2015;**4**:180-183

[2] Finkel T. Signal transduction by reactive oxygen species. Journal of Cell Biology. 2011;**194**(1):7-15

[3] Venegas C, García JA, Escames G, Ortiz F, López A, Doerrier C, et al. Extrapineal melatonin: Analysis of its subcellular distribution and daily fluctuations. Journal of Pineal Research. 2012;**52**(2):217-227

[4] World Health Organization (WHO). International Classification of Diseases. 11th Revision (ICD-11) ed. Geneva: WHO; 2018

[5] Ombelet W. WHO fact sheet on infertility gives hope to millions of infertile couples worldwide. Facts, Views & Vision in ObGyn. 2020;**12**(4):249-251

[6] Direkvand Moghaddam A, Delpisheh A, Sayehmiri K. An investigation of the worldwide prevalence of infertility as a systematic review. Qom University of Medical Sciences Journal. 2016;**10**(1):76-87

[7] Takeshima T, Usui K, Mori K, Asai T, Yasuda K, Kuroda S, et al. Oxidative stress and male infertility. Reproductive Medicine and Biology. 2021;**20**(1):41-52

[8] Garrido N, Meseguer M, Simon C, Pellicer A, Remohi J. Pro-oxidative and anti-oxidative imbalance in human semen and its relation with male fertility. Asian Journal of Andrology. 2004;**6**(1):59-66

[9] Zhaku V, Agarwal A, Beadini S, Henkel R, Finelli R, Beadini N, et al. Male infertility, oxidative stress and antioxidants. In: Vitamin E in Health and Disease-Interactions, Diseases and Health Aspects. London, UK: IntechOpen; 2021

[10] Gupta S, Ghulmiyyah J, Sharma R, Halabi J, Agarwal A. Power of proteomics in linking oxidative stress and female infertility. BioMed Research International. 2014;**2014**. Article ID: 916212. DOI: 10.1155/2014/916212

[11] Agarwal A, Aponte-Mellado A, Premkumar BJ, Shaman A, Gupta S. The effects of oxidative stress on female reproduction: A review. Reproductive Biology and Endocrinology. 2012;**10**(1):1-31

[12] Parul S, Preety G, Naveen K, Jaspreet K, Parminder K. Epigenetics and oxidative stress: A twin-edged sword in spermatogenesis. Andrologia. 2019;**00**:e13432

[13] Halliwell B, Gutteridge JM. Free Radicals in Biology and Medicine. USA: Oxford University Press; 2015

[14] Halliwell B. Reactive species and antioxidants. Redox biology is a fundamental theme of aerobic life. Plant Physiology. 2006;**141**:312-322. DOI: 10.1104/pp.106.077073

[15] Foyer CH, Noctor G. Redox homeostasis and antioxidant signaling: A metabolic interface between stress perception and physiological responses. The Plant Cell. 2005;**17**(7):1866-1875

[16] Bayr H. Reactive oxygen species. Critical Care Medicine. 2005;**33**(12):S498-S501

[17] Salvemini D, Cuzzocrea S. Oxidative stress in septic shock and

disseminated intravascular coagulation. Free Radical Biology and Medicine. 2002;**33**(9):1173-1185

[18] Mani S. Production of reactive oxygen species and its implication in human diseases. In: Free Radicals in Human Health and Disease. New Delhi: Springer; 2015. pp. 3-15

[19] Lambeth JD. NOX enzymes and the biology of reactive oxygen. Nature Reviews Immunology. 2004;**4**(3):181-189

[20] Nathan C, Cunningham-Bussel A. Beyond oxidative stress: An immunologist's guide to reactive oxygen species. Nature Reviews Immunology. 2013;**13**(5):349-361

[21] Turrens JF. Mitochondrial formation of reactive oxygen species. The Journal of Physiology. 2003;**552**(2):335-344

[22] Boveris A, Chance B. The mitochondrial generation of hydrogen peroxide. General properties and effect of hyperbaric oxygen. Biochemical Journal. 1973;**134**(3):707-716

[23] Turrens JF. Superoxide production by the mitochondrial respiratory chain. Bioscience Reports. 1997;**17**(1):3-8

[24] Bhandary B, Marahatta A, Kim HR, Chae HJ. An involvement of oxidative stress in endoplasmic reticulum stress and its associated diseases. International Journal of Molecular Sciences. 2012;**14**(1):434-456

[25] Santos CX, Tanaka LY, Wosniak J Jr, Laurindo FR. Mechanisms and implications of reactive oxygen species generation during the unfolded protein response: Roles of endoplasmic reticulum oxidoreductases, mitochondrial electron transport, and NADPH oxidase. Antioxidants & Redox Signaling. 2009;**11**(10):2409-2427

[26] Tu BP, Weissman JS. The FAD-and O2-dependent reaction cycle of Ero1-mediated oxidative protein folding in the endoplasmic reticulum. Molecular Cell. 2002;**10**(5):983-994

[27] Higa A, Chevet E. Redox signaling loops in the unfolded protein response. Cellular Signalling. 2012;**24**(8):1548-1555

[28] Angermüller S, Bruder G, Völkl A, Wesch H, Fahimi HD. Localization of xanthine oxidase in crystalline cores of peroxisomes. A cytochemical and biochemical study. European Journal of Cell Biology. 1987;**45**(1):137-144

[29] Fransen M, Nordgren M, Wang B, Apanasets O. Role of peroxisomes in ROS/RNS-metabolism: Implications for human disease. Biochimica et Biophysica Acta (BBA)-Molecular Basis of Disease. 2012;**1822**(9):1363-1373

[30] Zhang X, Rosenstein BS, Wang Y, Lebwohl M, Wei H. Identification of possible reactive oxygen species involved in ultraviolet radiation-induced oxidative DNA damage. Free Radical Biology and Medicine. 1997;**23**(7):980-985

[31] Winterbourn CC. Reconciling the chemistry and biology of reactive oxygen species. Nature Chemical Biology. 2008;**4**(5):278-286

[32] Halliwell B. Reactive oxygen species in living systems: Source, biochemistry, and role in human disease. The American Journal of Medicine. 1991;**91**(3):S14-S22

[33] Finkel T, Holbrook NJ. Oxidants, oxidative stress and the biology of ageing. Nature. 2000;**408**(6809):239-247

[34] Alfadda AA, Sallam RM. Reactive oxygen species in health and disease. Journal of Biomedicine and Biotechnology. 2012;**2012**:936486

[35] Liu Y, Zhao H, Li H, Kalyanaraman B, Nicolosi AC, Gutterman DD. Mitochondrial sources of H2O2 generation play a key role in flow-mediated dilation in human coronary resistance arteries. Circulation Research. 2003;**93**(6):573-580

[36] Go YM, Park H, Koval M, Orr M, Reed M, Liang Y, et al. A key role for mitochondria in endothelial signaling by plasma cysteine/cystine redox potential. Free Radical Biology and Medicine. 2010;**48**(2):275-283

[37] Guzy RD, Schumacker PT. Oxygen sensing by mitochondria at complex III: The paradox of increased reactive oxygen species during hypoxia. Experimental Physiology. 2006;**91**(5):807-819

[38] Dröge W. Free radicals in the physiological control of cell function. Physiological Reviews. 2002;**82**(1):47-95

[39] Merry TL, McConell GK. Do reactive oxygen species regulate skeletal muscle glucose uptake during contraction? Exercise and Sport Sciences Reviews. 2012;**40**(2):102-105

[40] Rajendran R, Garva R, Krstic-Demonacos M, Demonacos C. Sirtuins: Molecular traffic lights in the crossroad of oxidative stress, chromatin remodeling, and transcription. Journal of Biomedicine and Biotechnology. 2011;**2011**:368276

[41] Sandström ME, Zhang SJ, Bruton J, Silva JP, Reid MB, Westerblad H, et al. Role of reactive oxygen species in contraction-mediated glucose transport in mouse skeletal muscle. The Journal of Physiology. 2006;**575**(1):251-262

[42] Birben E, Sahiner UM, Sackesen C, Erzurum S, Kalayci O. Oxidative stress and antioxidant defense. World Allergy Organization Journal. 2012;**5**(1):9-19

[43] Lazo-de-la-Vega-Monroy ML, Fernández-Mejía C. Oxidative stress in diabetes mellitus and the role of vitamins with antioxidant action [chapter 9]. In: Morales-González JA, editor. Oxidative Stress and Chronic Degenerative Diseases – A Role for Antioxidants. Rijeka, Croatia: In-Tech; 2013

[44] Mirończuk-Chodakowska I, Witkowska AM, Zujko ME. Endogenous non-enzymatic antioxidants in the human body. Advances in Medical Sciences. 2018;**63**(1):68-78

[45] Wu JQ, Kosten TR, Zhang XY. Free radicals, antioxidant defense systems, and schizophrenia. Progress in Neuro-Psychopharmacology and Biological Psychiatry. 2013;**46**:200-206

[46] Shahidi F, Zhong Y. Novel antioxidants in food quality preservation and health promotion. European Journal of Lipid Science and Technology. 2010;**112**(9):930-940

[47] Babula P, Masarik M, Adam V, Eckschlager T, Stiborova M, Trnkova L, et al. Mammalian metallothioneins: Properties and functions. Metallomics. 2012;**4**(8):739-750

[48] Agarwal A, Prabakaran SA. Mechanism,measurement,andprevention of oxidative stress in male reproductive physiology. 2005;**43**(11):963-974

[49] Gil-Guzman E, Ollero M, Lopez MC, Sharma RK, Alvarez JG, Thomas AJ Jr, et al. Differential production of reactive oxygen species by subsets of human spermatozoa at different stages of maturation. Human Reproduction. 2001;**16**(9):1922-1930

[50] Hermo L, Pelletier RM, Cyr DG, Smith CE. Surfing the wave, cycle, life

history, and genes/proteins expressed by testicular germ cells. Part 1: Background to spermatogenesis, spermatogonia, and spermatocytes. Microscopy Research and Technique. 2010;**73**(4):241-278

[51] Badouard C, Ménézo Y, Panteix G, Ravanat JL, Douki T, Cadet J, et al. Determination of new types of DNA lesions in human sperm. Zygote. 2008;**16**(1):9-13

[52] Laqqan M, Solomayer EF, Hammadeh M. Aberrations in sperm DNA methylation patterns are associated with abnormalities in semen parameters of subfertile males. Reproductive Biology. 2017;**17**(3):246-251

[53] Sharma U, Conine CC, Shea JM, Boskovic A, Derr AG, Bing XY, et al. Biogenesis and function of tRNA fragments during sperm maturation and fertilization in mammals. Science. 2016;**351**(6271):391-396

[54] Asadi N, Bahmani M, Kheradmand A, Rafieian-Kopaei M. The impact of oxidative stress on testicular function and the role of antioxidants in improving it: A review. Journal of Clinical and Diagnostic Research: JCDR. 2017;**11**(5):IE01

[55] Saleh RA, HCLD AA. Oxidative stress and male infertility: From research bench to clinical practice. Journal of Andrology. 2002;**23**(6):737-752

[56] Walczak-Jedrzejowska R, Wolski JK, Slowikowska-Hilczer J. The role of oxidative stress and antioxidants in male fertility. Central European Journal of Urology. 2013;**66**(1):60

[57] Fraczek M, Kurpisz M. The redox system in human semen and peroxidative damage of spermatozoa. Postepy Higieny i Medycyny Doswiadczalnej (Online). 2005;**59**:523-534

[58] Gałecka E, Jacewicz R, Mrowicka M, Florkowski A, Gałecki P. Antioxidative enzymes--structure, properties, functions. Polski Merkuriusz Lekarski: Organ Polskiego Towarzystwa Lekarskiego. 2008;**25**(147):266-268

[59] Alahmar AT. Role of oxidative stress in male infertility: An updated review. Journal of Human Reproductive Sciences. 2019;**12**(1):4

[60] Wagner H, Cheng JW, Ko EY. Role of reactive oxygen species in male infertility: An updated review of literature. Arab Journal of Urology. 2018;**16**(1):35-43

[61] Mayorga-Torres BJ, Camargo M, Cadavid AP, Du Plessis SS, Cardona Maya WD. Are oxidative stress markers associated with unexplained male infertility? Andrologia. 2017;**49**(5):e12659

[62] Zirkin BR, Chen H. Regulation of Leydig cell steroidogenic function during aging. Biology of Reproduction. 2000;**63**(4):977-981

[63] Kaur P, Bansal MP. Effect of experimental oxidative stress on steroidogenesis and DNA damage in mouse testis. Journal of Biomedical Science. 2004;**11**(3):391-397

[64] Stocco DM, Clark BJ. Role of the steroidogenic acute regulatory protein (StAR) in steroidogenesis. Biochemical Pharmacology. 1996;**51**(3):197-205

[65] Agarwal A, Saleh RA, Bedaiwy MA. Role of reactive oxygen species in the pathophysiology of human reproduction. Fertility and Sterility. 2003;**79**(4):829-843

[66] Ferrini MG, Gonzalez-Cadavid NF, Rajfer J. Aging related erectile dysfunction—potential mechanism to

halt or delay its onset. Translational Andrology and Urology. 2017;**6**(1):20

[67] Gratzke C, Angulo J, Chitaley K, Dai Y, Kim NN, Paick JS, et al. Anatomy, physiology, and pathophysiology of erectile dysfunction. The Journal of Sexual Medicine. 2010;**7**:44575

[68] Paolicchi A, Pezzini A, Saviozzi M, Piaggi S, Andreuccetti M, Chieli E, et al. Localization of a GSH-dependent dehydroascorbate reductase in rat tissues and subcellular fractions. Archives of Biochemistry and Biophysics. 1996;**333**(2):489-495

[69] Özkan KU, Boran C, Kilinc M, Garipardic M, Kurutaş EB. The effect of zinc aspartate pretreatment on ischemia-reperfusion injury and early changes of blood and tissue antioxidant enzyme activities after unilateral testicular torsion-detorsion. Journal of Pediatric Surgery. 2004;**39**(1):91-95

[70] Verma RJ, Nair A. Ameliorative effect of vitamin E on aflatoxin-induced lipid peroxidation in the testis of mice. Asian Journal of Andrology. 2001;**3**(3):217

[71] Bray TM, Bettger WJ. The physiological role of zinc as an antioxidant. Free Radical Biology and Medicine. 1990;**8**(3):281-291

[72] Agarwal A, Gupta S, Sharma RK. Role of oxidative stress in female reproduction. Reproductive Biology and Endocrinology. 2005;**3**(1):1-21

[73] Hanukoglu I. Antioxidant protective mechanisms against reactive oxygen species (ROS) generated by mitochondrial P450 systems in steroidogenic cells. Drug metabolism Reviews. 2006;**38**(1-2):171-196

[74] Bernal AB, Vickers MH, Hampton MB, Poynton RA, Sloboda DM. Maternal undernutrition significantly impacts ovarian follicle number and increases ovarian oxidative stress in adult rat offspring. PLoS One. 2010;**5**(12):e15558

[75] Revelli A, Piane LD, Casano S, Molinari E, Massobrio M, Rinaudo P. Follicular fluid content and oocyte quality: From single biochemical markers to metabolomics. Reproductive Biology and Endocrinology. 2009;**7**(1):1-3

[76] Behl R, Pandey RS. FSH induced stimulation of catalase activity in goat granulosa cells in vitro. Animal Reproduction Science. 2002;**70**(3-4):215-221

[77] Bausenwein J, Serke H, Eberle K, Hirrlinger J, Jogschies P, Hmeidan FA, et al. Elevated levels of oxidized low-density lipoprotein and of catalase activity in follicular fluid of obese women. MHR: Basic Science of Reproductive Medicine. 2009;**16**(2):117-124

[78] Borek C. Antioxidants and radiation therapy. The Journal of Nutrition. 2004;**134**(11):3207S-3209S

[79] Said RS, Nada AS, El-Demerdash E. Sodium selenite improves folliculogenesis in radiation-induced ovarian failure: A mechanistic approach. PLoS One. 2012;**7**(12):e50928

[80] Bedaiwy MA, Falcone T, Mohamed MS, Aleem AA, Sharma RK, Worley SE, et al. Differential growth of human embryos in vitro: Role of reactive oxygen species. Fertility and Sterility. 2004;**82**(3):593-600

[81] Jančar N, Kopitar AN, Ihan A, Klun IV, Bokal EV. Effect of apoptosis and reactive oxygen species production in human granulosa cells on oocyte

fertilization and blastocyst development. Journal of Assisted Reproduction and Genetics. 2007;**24**(2):91-97

[82] Behrman HR, Kodaman PH, Preston SL, Gao S. Oxidative stress and the ovary. Journal of the Society for Gynecologic Investigation. 2001;**8**(1_suppl):S40-S42

[83] Cigliano L, Balestrieri M, Spagnuolo MS, Dale B, Abrescia P. Lecithin-cholesterol acyltransferase activity during maturation of human preovulatory follicles with different concentrations of ascorbate, a-tocopherol and nitrotyrosine. Reproduction, Fertility and Development. 2002;**14**(1):15-21

[84] Cruz MH, Leal CL, Cruz JF, Tan DX, Reiter RJ. Essential actions of melatonin in protecting the ovary from oxidative damage. Theriogenology. 2014;**82**(7):925-932

[85] Taketani T, Tamura H, Takasaki A, Lee L, Kizuka F, Tamura I, et al. Protective role of melatonin in progesterone production by human luteal cells. Journal of Pineal Research. 2011;**51**(2):207-213

[86] Espey LL, Stein VI, Dumitrescu J. Survey of antiinflammatory agents and related drugs as inhibitors of ovulation in the rabbit. Fertility and Sterility. 1982;**38**(2):238-247

[87] Brännström M, Mayrhofer G, Robertson SA. Localization of leukocyte subsets in the rat ovary during the periovulatory period. Biology of Reproduction. 1993;**48**(2):277-286

[88] Kodaman PH, Behrman HR. Endocrine-regulated and protein kinase C-dependent generation of superoxide by rat preovulatory follicles. Endocrinology. 2001;**142**(2):687-693

[89] Agarwal A, Prabakaran S, Allamaneni SS. Relationship between oxidative stress, varicocele and infertility: A meta-analysis. Reproductive BioMedicine Online. 2006;**12**(5):630-633

[90] Attaran M, Pasqualotto E, Falcone T, Goldberg JM, Miller KF, Agarwal A, et al. The effect of follicular fluid reactive oxygen species on the outcome of in vitro fertilization. International Journal of Fertility and Women's Medicine. 2000;**45**(5):314-320

[91] Paszkowski T, Clarke RN, Hornstein MD. Smoking induces oxidative stress inside the Graafian follicle. Human Reproduction. 2002;**17**(4):921-925

[92] Oyawoye O, Abdel Gadir A, Garner A, Constantinovici N, Perrett C, Hardiman P. Antioxidants and reactive oxygen species in follicular fluid of women undergoing IVF: Relationship to outcome. Human Reproduction. 2003;**18**(11):2270-2274

[93] Pasqualotto EB, Agarwal A, Sharma RK, Izzo VM, Pinotti JA, Joshi NJ, et al. Effect of oxidative stress in follicular fluid on the outcome of assisted reproductive procedures. Fertility and Sterility. 2004;**81**(4):973-976

[94] Lázár L. The role of oxidative stress in female reproduction and pregnancy. Oxidative Stress In Disease. 2012;**25**:313-336

[95] Agarwal A, Sengupta P, Durairajanayagam D. Role of L-carnitine in female infertility. Reproductive Biology and Endocrinology. 2018;**16**(1):1-8

[96] McDowell LR, Wilkinson N, Madison R, Felix T. Vitamins and minerals functioning as antioxidants with supplementation

considerations. In: Florida Ruminant Nutrition Symposium. Vol. 352. Gainesville, FL, USA: Best Western Gateway Grand; 2007. pp. 1-17

[97] Sekhon L, Gupta S, Kim Y, Agarwal A. Female infertility and antioxidants. Current Women's Health Reviews. 2010;**6**(2):84-95

Section 2

Antioxidants

Chapter 9

Antioxidant Phytochemicals as Novel Therapeutic Strategies against Drug-Resistant Bacteria

Bhavana Gangwar, Santosh Kumar and Mahendra P. Darokar

Abstract

The antibiotic resistance in pathogenic bacteria is a major concern and the emergence of novel multidrug-resistant (MDR) strains are a growing threat worldwide. Bacterial resistance to antibiotics has become a serious problem of public health that concerns almost all antibacterial agents and that manifests in all fields of their application. Therefore, novel antimicrobial compounds against new bacterial targets and drug resistance mechanisms are urgently needed. Plants are well-known sources of structurally diverse phytochemicals such as alkaloids, flavonoids, phenolics, and terpenes, which plays important roles in human health. Plant-derived antimicrobial agents are an attractive and ongoing source of new therapeutics. Natural compounds that prevent and treat infections through dual action mechanisms such as oxidative stress against pathogens and antioxidant action in the host cell hold promising potential for developing novel therapeutics. Identification of detailed mechanisms of action of such phytomolecules with both antioxidant and antimicrobial activities may help to develop novel antimicrobial therapeutics and benefit overall human health. The purpose of this chapter is to summarize important antioxidant phytochemicals, and focusing on their potential role in the management of drug-resistant bacterial infections.

Keywords: antioxidant, drug resistance, oxidative stress, phytochemicals, drug-resistant bacteria

1. Introduction

Antimicrobial resistance has now become a serious public health issue worldwide. Resistance to antimicrobials is a growing challenge that limits treatment options against serious pathogens and therefore new effective treatment strategies are needed [1]. Infections caused by *Staphylococcus aureus, Streptococcus pyogenes, Streptococcus pneumoniae, Enterococcus faecalis*, and Gram-negative bacteria such as *Escherichia coli, Klebsiella pneumonia, Salmonella typhi, Pseudomonas aeruginosa*, are among the most common bacteria that have developed drug-resistant to many antibiotics. According to the Centers for Disease Control and Prevention (CDC) 2019 AR threats report [2] penicillin-resistant *Pneumococcus*, drug-resistant *Campylobacter* sp., methicillin-resistant *Staphylococcus aureus* (MRSA), vancomycin-resistant *Enterococcus faecalis* (VRE), multidrug-resistant of *Pseudomonas aeruginosa*,

Salmonella typhi, *Shigella sp.*, and *Mycobacterium tuberculosis* (MDR-TB) are serious threat and major causes of worldwide outbreaks of both the community infections and hospitals [3]. While carbapenem-resistant *Enterobacterales,* drug-resistant *Gonorrhea* and *Clostridioides difficile* are grouped among urgent treats [3]. The spread of multidrug-resistant (MDR) strains of pathogenic bacteria necessitate the discovery and deployment of new classes of antibacterial and compounds that can combat resistant strains and the spread of drug resistance.

During infections in humans, immune cells produce reactive oxygen and nitrogen species (RONS) which is used as part of warfare activity against pathogens [4, 5]. Oxidative stress induced by intracellular bacterial infection or other metabolic processes can cause inflammation and cellular damage however RONS production helps to kill bacterial pathogens. Unfortunately, pathogens have evolved a number of adaptive mechanisms against host-mediated defense systems and RONS [6, 7]. Microorganisms' survival strategies against RONS include the expression of various enzymes catabolizing RONS such as catalase, peroxidases, and biofilm formation also helps pathogens overcome the immune defense system [8, 9]. Therefore, targeting bacterial redox systems could present an important tool to combat such infections. Indeed, significant progress has been made in identifying several natural source of antioxidants that may also cause oxidative stress as a part of the antibacterial mechanism of action [10–12]. This emerging field needs further focus on the redox biology of antioxidants with antimicrobial activity by oxidative stress to cure intracellular bacterial pathogens.

Plants have an exceptional ability to produce cytotoxic agents to protect themselves from pathogenic microbes in their environment. Plant-derived secondary metabolites with antibacterial properties can be a source for designing novel therapeutics [13–15]. Historically, traditional medicines based on plants have made a considerable amount of contributions to human health. Plants are rich in a wide variety of secondary metabolites such as terpenoids, alkaloids, polyphenols, and tannins with a diverse set of biological activities. During the last decades, there is increasing interest to explore ancient remedies. A significant number of works, e.g., biological screening, isolation as well as clinical trials have been done for a variety of plants to unlock the secrets of herbal remedies [15]. Antimicrobials with reuse potential, which can be used in combination with drug treatments against drug-resistant pathogens, are identified using this approach. For example, plants derived alkaloids such as tomatidine and berberine are reported to be highly effective against drug-resistant microbes and also show synergy with antibiotics against *S. aureus,* and *E. coli* [16]. Therefore, screening and identification of compounds responsible for antimicrobial activity can be the foundation of a novel class of drugs. In this chapter, we have summarized the importance of medicinal and aromatic plants in the management of drug-resistant bacterial pathogens.

2. Role of oxidative stress and antioxidants mechanisms in health and infectious diseases

The paradox of oxidative stress is that it plays a dual role in the disease and health of humans. The importance of oxidative stress mechanisms in living cells is based on a balance between oxidants and antioxidants [17]. During metabolic reactions and infection, various types of RONS are produced in human by enzymes like myeloperoxidase, oxidases, and nitric oxide synthase, however, excess of these RONS can also damage host tissues and thus are reduced by the human cellular antioxidative defense system that includes enzymes like peroxidases, superoxide dismutase (SOD), catalase (CAT), glutathione (GSH), and others which eliminates the excess of reactive oxygen species (ROS) such as hydroxyl radicals ($OH^{\cdot}$), superoxide anions ($O2^{\cdot -}$), alkoxyl radicals ($RO^{\cdot}$), and peroxy radicals ($ROO^{\cdot}$). As a result, we require antioxidant supplements like vitamin C, vitamin E, carotenoids, and polyphenols to avoid oxidative stress [18, 19]. Supplementing a low amount of oxidative stress may also help to signal processes to express enzymes that detoxify the RONS [20]. This means low level of physiological oxidative stress can be beneficial to counteract excess oxidants produced during stress and infections that may result in cellular damage.

During phagocytosis, NADH-dependent oxidase (NOX) mediated burst of superoxide anion ($O2^{-}$) causes bactericidal oxidative stress [8, 21, 22]. This free radical is converted to hydrogen peroxide (H_2O_2) by superoxide dismutases. Infection control requires the presence of the NOX protein and abnormalities in the genes that produce the NOX protein makes it more vulnerable to bacterial and fungal infections [8, 23, 24]. H_2O_2 produced during phagocytosis passes through bacterial membranes barrier and interacts with ferrous iron (Fe^{2+}) and thiol groups (-SH) of the cysteine in proteins which can eventually inactivate the function of enzymes [25]. During the Fenton reaction, H_2O_2 oxidizes Fe^{2+} to Fe^{3+} and produces hydroxyl radicals (OH^{-}), which further damage bacterial DNA, proteins, and lipids [22, 26]. Myeloperoxidases expressed in macrophages and neutrophils produce hypochlorous acid (HClO) from the reaction between H_2O_2 and chloride ion (Cl^{-})· HClO has a stronger antibacterial effect than H_2O_2 [21, 22]. Later phases of phagocytosis activate inducible nitric oxide synthases (iNOS). Nitric oxide ($NO^{\bullet}$) is produced by these enzymes from L-arginine. Peroxynitrite is formed when nitric oxide reacts with the superoxide ion produced by NOX proteins. Peroxynitrite can directly oxidize the thiol groups of sulfur-containing amino acids, or it can break down into nitrogen dioxide and hydroxyl radicals, which can damage the sulfur-containing proteins in bacteria [26, 27].

3. Bacterial antioxidant and redox pathway mechanisms

As described above, during phagocytosis bacteria are exposed to several RONS but they can still be growing in the intracellular environment under these oxidative conditions [8, 22, 26, 28, 29]. It is important to note that microorganism also possesses several enzymes like catalase, peroxidases, and superoxide dismutase. These protect against oxidative stress with a complex set of enzymatic activities (**Figure 1**) that can be divided into two categories: (i) preventive mechanisms, which are based on protein scavengers to inactivate RONS, and (ii) the repair mechanism, which is based on the reduction of the thiol groups of oxidized protein to restore enzyme activity [8]

The expression of antioxidant enzymes is controlled by transcriptional regulators that can interact with RONS based on thiol switches or metal centers. For example, OxyR is redox regulator that is reported to act as transcriptional activators or repressors in several bacteria [30]. Homologs of the MarR-family, thiol-based transcriptional regulators e.g. the sodium hypochlorite sensor (HypS) are found in the genomes of a variety of pathogens. Oxidation the thiol groups by RONS causes conformational changes in transcriptional regulators and modulates their binding capacity to promoters of genes encoding scavenger enzymes such catalases (Kat), superoxide dismutases, glutathione peroxidases (GPx), and peroxiredoxins (Prx). Many pathogens such as *Mycobacterium tuberculosis* are known to use combination

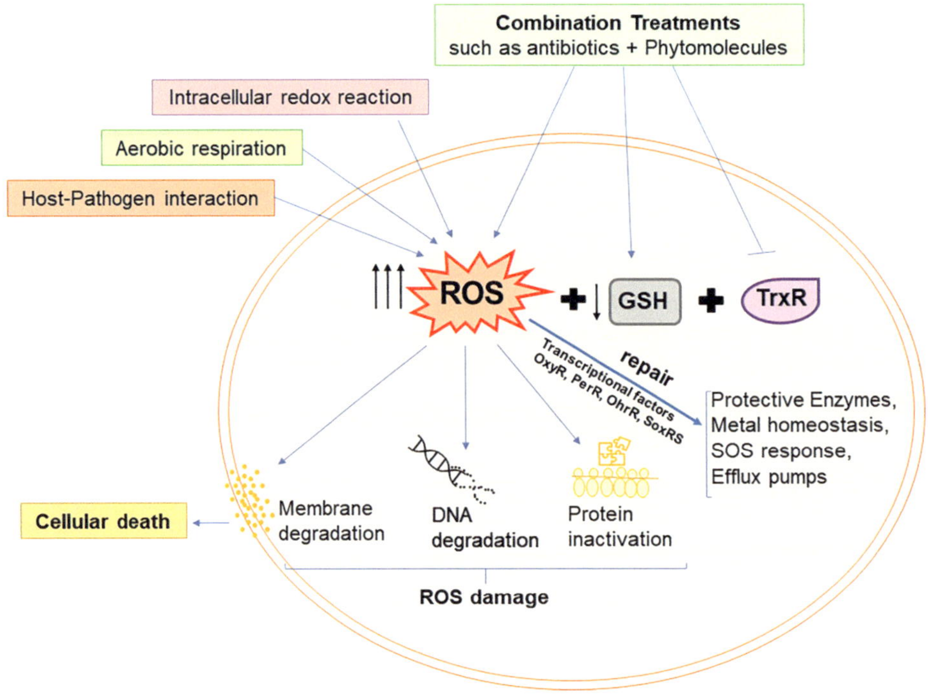

Figure 1.
Overview of oxidative stress and response mechanism in bacteria. Oxidative stress is caused by the accumulation of reactive oxygen species (ROS) caused by both exogenous and endogenous sources. Bacterial cells are damaged by ROS because it causes DNA degradation, protein inactivation, membrane degradation, etc. Bacteria use a variety of mechanisms such as repair mechanisms and protective enzyme synthesis, metal homeostasis, the SOS response, and efflux pumps to counteract oxidative stress. OxyR, PerR, OhrR, and SoxRS are transcriptional factors that control the expression of oxidative stress response in bacteria.

of these enzymes to overcome RONS challenges. Loss of one or more of these genes directly affects resistance to RONS and survival of bacteria [31, 32]. Extracellular thioredoxins (Etrx) found on the bacterial surface, have been found in human pathogens such as *M. tuberculosis*, *S. pneumoniae*, *N. gonorrhoeae*, as well as in plant-associated bacteria such as *Agrobacterium tumefaciens*, and *Bradyrhizobium japonicum* [8, 33, 34]. Although the targets of Etrx proteins are unknown, deletion of the genes encoding Etrx proteins reduced the pathogenicity of *M. tuberculosis* and *S. pneumoniae* [33]. However, more research is needed to fully understand the role of such surfaceome-associated proteins and their involvement during infection.

4. Plant as potential source of preventative and therapeutic agents of oxidative stress and disease

Medicinal plants and their extracted phytochemicals are widely used in the treatment of a variety of diseases, including bacterial, fungal, viral and cancer, as well as oxidative stress-related problems [35–37]. Due to the anti-oxidative, anti-inflammatory, anti-microbial, and wound-healing characteristics natural phytochemicals (**Figure 2**) reduce the risk of diseases [9]. These substances have been studied extensively to facilitate their application as phytomedicine in the pharmaceutical field. Several phytomolecules such as quercetin prevents oxidative damage in human

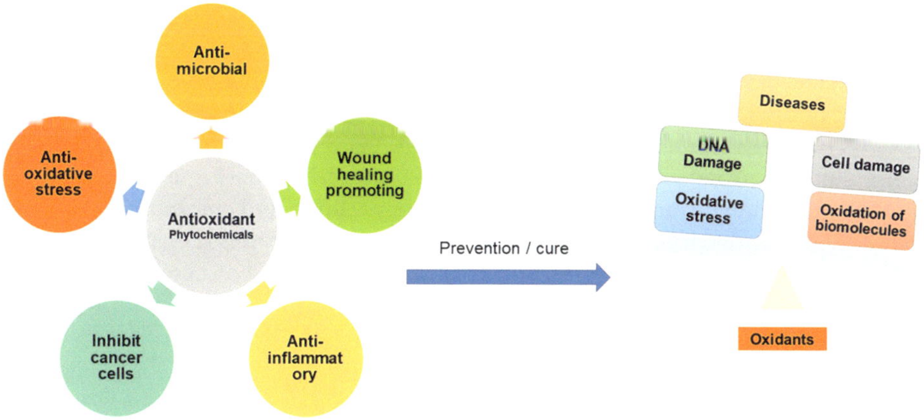

Figure 2.
Schematic representation of the potential roles of antioxidant phytochemicals.

by influencing glutathione levels, enzymes, signal transduction pathways, and ROS production as well as show antibiofilm activity and bacteriostatic properties against several pathogens like *E. coli, S. aureus* and *P. aeruginosa* by means of promoting oxidative cellular stress targeting a wide range of cellular component [12, 38, 39]

Natural products have shown to be a never-ending source of novel medicines. Medicinal plants have been used in drug development since ancient times, and they continue to provide novel and important roles against a variety of medicinal targets, including infections, cancer, HIV/AIDS, Alzheimer's disease, and malaria [40]. Ayurveda and Charaka Samhita have contributed significantly to the discovery of new drugs and chemical entities due to India's abundant biodiversity and traditional medicinal herbs [41]. Ancient Chinese and African traditional medicine has a long history of medicinal plants for several physiological conditions [42]. Early pharmaceuticals such as cocaine, codeine, digitoxin, and quinine, as well as morphine, were discovered from medicinal plants, and are still used today [43]. Natural compounds or chemicals inspired by nature provide more than 80% of all medicines used today [44]. On the other hand, drug discovery from natural plants is a time-consuming and laborious process.

To date, several bioactive chemicals have been identified and described from medicinal plants that have been successfully exploited for biomedical purposes. There are more than 100 natural product-derived compounds already in clinical studies [44]. Traditional medicinal systems have a long history, dating back more than 60,000 years [45, 46]. Herbal remedies are used in all developed countries, and the WHO estimates that approximately 75% of the world's population uses medicinal items as an alternative to allopathic medicines [47]. It is interesting to note that medicinal plants provide 35% of all drugs recommended and prescribed today [47, 48]. Natural source phytomedicine in pure is either used directly or converted using appropriate chemical or microbiological processes until used as medicine [37]. Alkaloids, phenolics, and terpenes are examples of natural chemicals that have been demonstrated to be effective against many pathogens. Pomolic acid, oleanolic acid [49], gallic acid, chebulagic acid, and other galloyl glucose [50] have also been reported to inhibit HIV integrase. Researchers continue to uncover phytochemicals in many plants that are important for drug development; Their actions may offer new hope in treating various infections or diseases as well as reducing toxicity.

5. Classification of phytochemicals

Natural bioactive compounds or phytochemicals have significant physiological effects on human health. They are a major source of diver's chemicals, making them a potential source of new drugs. Phytochemicals are classified as primary and secondary constituents according to their importance in plant metabolism (**Figure 3**). Generally, primary metabolite includes carbohydrates, proteins, lipid, and chlorophyll; the secondary metabolite has been classified into six major categories (i.e., phenolic, coumarins, glycoside, chalcones, terpenes, and alkaloid) based on chemical structures and characteristics.

5.1 Alkaloids

Alkaloids are produced by plants as protective agents against attack by predators. Alkaloids are basic compounds that contain heterocyclic nitrogen atoms that react with acids to produce salts. Most of all alkaloids are sour in taste. Morphine, the first alkaloid discovered in *Papaver somniferum*, has antibacterial properties [51]. Antimicrobial activities were discovered in diterpene alkaloids from Ranunculaceae plants [52]. Berberine is an isoquinoline alkaloid isolated from berberine species that has antibacterial and antiviral properties [53] Quinine, an alkaloid, was the first successful antimalarial drug extracted from the Cinchona tree. Atropine, codeine, coniine, caffeine, hyoscyamine, scopolamine, sanguinarine, etc., are the other examples of alkaloid found in nature. Alkaloids have diverse physiological effects and have been reported to possess antibacterial, anesthetic, antiinflammatory, antimitotic, analgesic, hypnotic, psychotropic, and antitumor activity, and many others.

5.2 Flavonoids

Flavonoids are polyphenolic chemicals found in vascular plants that come in the form of aglycones, glucosides, and methylated derivatives. Tomatoes, grapes, berries, apples, onions, kale, and lettuce are rich sources of flavonoids. Flavonoids are divided into two groups: flavone and isoflavone depending on the position of the benzenoid substituent. The majority of flavonoids are found in the conjugated form

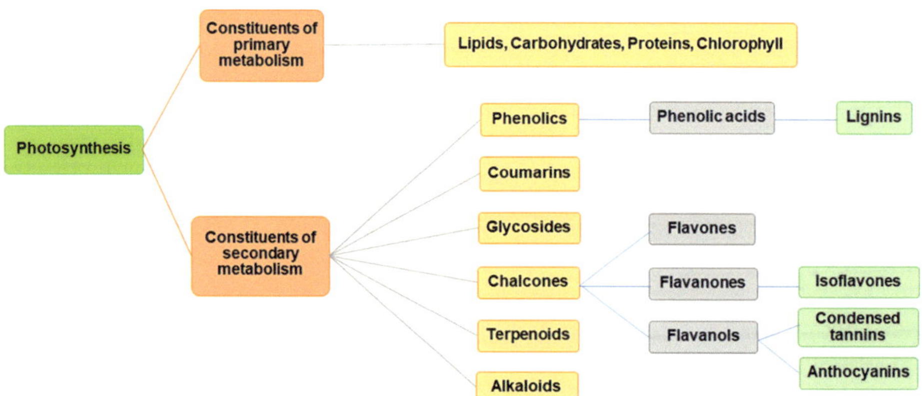

Figure 3.
Phytochemical classification demonstrating the link between primary and secondary metabolism in plants.

in nature and can be classified as monoglycosidic, diglycosidic, or polyglycosidic within each class. The carbohydrate unit might be L-rhamnose, D-glucose, galactose, or arabinose, and the glycosidic linkage is usually found at position 3 or 7 [37]. Plants, animals, human, and microorganisms all use flavonoids for a range of biological functions. Flavonoids have been linked to improved human health, and they are currently being studied for antibacterial activity and chemoprevention [54, 55]. Apigenin, quercetin, kaempferol, fisetin, glabridin, and myricetin are most studied flavonoids. Flavonoids act as antioxidants and also insilico modeling and docking studies suggest potential as an antibiocbial activity. Apigenin, and quercetin are known to show antibacterial activity [55]. Quercetin and its analog penta-O-ethylquercetin were found to be potential inhibitors of New Delhi metallo-β-lactamase-1 (NDM-1). Several researches have reported synergy between flavonoids and antibiotics against resistant strains of bacteria [55, 56].

5.3 Phenolics and polyphenols

Phenolic acids contain carboxylic acid functional group. The hydroxycinnamic and hydroxybenzoic structures are found in naturally occurring phenolic acids [57]. Phenolic compounds possess strong anti-inflammatory, antioxidant, and antimicrobial activities [57]. Plant phenolic compounds and phenolic compound-rich herbal extracts control cell proliferation, survival, and apoptosis via modulating the amounts of reactive oxygen species (ROS) in cells. Recent research has also shown that phenolic compounds undergo change in the gut microbiota, gaining new characteristics that enhance their biological activity [57]. Tennins are polyphenolic chemicals that can form complexes with nucleic acids, proteins, and polysaccharides among other elements [37]. Phenolic compounds such as catechins, epigallocatechin gallate, galangin are reviewed for their antibacterial activity [58]. Ferulic acid, coumaric acid, chlorogenic acid, and caffeic acid have shown efficacy against *S. aureus* [59, 60].

5.4 Terpenes

Terpenes are natural occurring chemicals found in plants and are known for the aromas and flavors. Terpene contains isoprene units made up of five carbons atoms. Terpene chemical formula is $(C_5H_8)_n$ and their hydrocarbons are characterized by number of isoprene units [61]. Terpenes are important for plant growth and development, physiological processes, and response to the environment. Cannabis is important terpenes known for several uses such as aroma and taste. Terpenoids are essential oils and volatile chemicals found in higher medicinal plants. Terpenoids also show antimicrobial activity [61]. Monoterpenes such as menthol, sabinene, limonene, and carvone have shown strong antibacterial activity against *S. aureus* [61, 62]. Sesquiterpene (Patchouli alcohol), Diterpene (Artemisinin and Andrographolide), and Triterpene (Oleanolic acid) also show antibacterial activity in bacterial strains [61, 63–65].

6. Antioxidant phytochemicals: mechanism of action

Antioxidants are a type of defense mechanism that protects the human body from oxidative damage caused by free radicals. There are many different types of naturally occurring antioxidants with different physical and chemical properties, processes, and mechanisms of action. Antioxidant activity has been reported in medicinal plants rich in vitamins, carotenoids, flavonoids, polyphenols, and anthocyanins. The antioxidant mechanism of phytochemicals has been reviewed in detail [66]. Briefly, phytochemicals exert antioxidant activity by the following means; 1. Free radical scavenging activity; antioxidants are known to chelate free radicals, transition metals, and remove electrons or hydrogen from substance. 2. Inhibition of the expressions of free radical creating enzymes or induce the expressions of other antioxidant enzymes, e.g., modulation of host nuclear factor erythroid 2 (Nrf2), a master regulator of antioxidant defense system, that controls over a dozen of enzymes such as glutathione S-transferases (GSTs) and NAD(P)H: quinone oxidoreductase 1 (NQO1) [66]. Resveratrol, anthocyanins, and curcumin are phytochemicals that modulate prostaglandin formation and Nrf2 activity, inhibit enzymes, and increase cytokine production, all of which may help reduce inflammation [67, 68]. 3. Prevents lipid peroxidation, DNA damage, and protein modification caused by ROS [66].

7. Anti-microbial phytochemicals: mechanism of action

The use of medicinal plants and their extracts for the treatment of all infectious ailments was widespread. In recent years, many research works have been conducted on the efficacy of plant phytochemicals as antibacterial agents *in vitro* and *in vivo* [16, 69]. Some of the important phytochemicals found active against Gram-negative and Gram-positive bacteria with their mode of action are listed in **Table 1**. Different antioxidant phytochemicals show various modes of action against pathogens such as inhibition of: (1) efflux pump, (2) DNA gyrase, (3) protein synthesis, (4) cell division and metabolic enzyme, (5) cell wall and cell membrane, (6) energy production, and (7) Biofilm. Here we focus more on oxidative stress as a new therapeutic strategy with the goal of unbalancing the redox defenses of bacterial pathogens. The intervention of redox homeostasis is becoming a potential target for combating drug resistance in bacteria. RONS-generating plant-derived antimicrobials have received a lot of attention in recent years [8]. Because of their ability to generate dose-dependent oxidative shifts in the bacteria, some plant-derived chemicals have been shown to have antibacterial activity. Several pathways have been discovered using system biology approaches to disrupt the antioxidant systems of bacterial pathogens [70]. Ebselen (also known as PZ 51, DR3305, and SPI-1005) is reported for antioxidant activity in humans, however, in bacteria such as *M. tuberculosis* or *S. aureus,* it can inhibit growth by causing oxidative stress [71]. Allicin, a defensive molecule, produced by garlic (*Allium sativum*) is the most investigated. Allicin can oxidize proteins' thiol groups in a dose-dependent manner. In *S. aureus* and *Bacillus subtilis*, allicin's antibacterial efficacy and oxidative role have both been proven. Allicin causes high disulfide stress in these bacteria, lowering their viability considerably. Curcumin, a strong antioxidant substance found in turmeric is reported as an antibacterial compound which disrupts bacterial quorum sensing system, cell wall and cell membrane, biofilm and virulence gene expression, and also shows synergy with antibiotics [72]. Recent published work on glabridin also showed dose-dependent activity as antioxidant, antibacterial as well as anti-biofilm against multidrug-resistant *S. aureus* [73, 74].

Some conventional antibiotics such as Norfloxacin, Kanamycin, Rifampicin, and Quinones can also generate RONS as part of their mechanism of action [8, 75]. Combining RONS-generating antimicrobials with antibiotics may have a synergistic effect against specific bacterial infections. Similarly, combining antimicrobials with

Phytochemical class	Phytochemicals	Source	Mode of action	Reported pathogens	Reference
Alkaloid	Lysergol	*Ipomoea muricata*	Efflux pump inhibitor	*E. coli*	[76]
	Reserpine	*Rauwolfia serpentina*	Efflux pump inhibitor	*Streptococcus sp., Staphylococcus sp.,*	[77]
	Berberine	*Berberis species*	Cell division/Protein/DNA synthesis inhibitor	*E. coli*	[78]
	Tomatidine	*Solanaceous plants*	ATP synthase inhibitor	*Staphylococcus spp., Listeria sp., Bacillus sp.*	[79]
	Matrine	*Thermopsis lanceolata R. Brown*	Protein synthesis inhibitor	*E. coli, E. aerogenes, P. vulgaris, S. epidermidis, B. subtilis*	[80, 81]
	Lycorine	*Pancratium Foetidum Pom*	Protein synthesis inhibitor	*E. coli, S. aureus, P. aerugin, E. cloac*	[82]
	Chabamide	*Piper chaba*	Protein synthesis inhibitor	*M. tuberculosis*	[83]
Flavonoids	Apigenin	*Petroselinum crispum*	DNA gyrase inhibitor	*P. aeruginosa, L. monocytogenes, Aeromonas hydrophila,*	[84–86]
	Quercetin	*Vaccinium* sect. *Cyanococcus*	DNA gyrase inhibitor	*P. aeruginosa*	[87]
	Kaempferol	*Brassica oleracea var. italica*	DNA gyrase inhibitor	*E. coli*	[88]
	Myricetin	*Vaccinium Oxycoccus (Cranberry)*	DNA gyrase inhibitor	*E. coli, S. aureus*	[89]
	Galangin	*Robinia pseudoacacia*	DNA gyrase inhibitor	*S. aureus*	[90]
	Glabridin	*Glycyrrhiza glabra*	Oxidative stress inducing and protein degradation, Biofilm inhibition by modulation of cell surfaceome	*Methicillin-resistant S. aureus*	[73, 74]

Phytochemical class	Phytochemicals	Source	Mode of action	Reported pathogens	Reference
	Baicalin	*Scutellaria baicalensis*	Enzymes inhibition, protein kinase inhibition, Biofilm inhibition	*Methicillin-resistant S. Aureus, P. aeruginosa, P. gingivalis, M. tuberculosis, Streptococcus mutans, Fusobacterium nucleatum, Aggregatibacter actinomycetemcomitans*	[91, 92]
Phenolics and polyphenols	Taxifolin	*Allium cepa*	Beta-Ketoacyl acyl carrier protein synthase inhibitor	*Enterococcus faecalis*	[93]
	Baicalein	*Scutellaria baicalensis*	Efflux pump inhibitor	*Methicillin-resistant S. aureus*	[94]
	Resveratrol	*Vitis vinifera*	Efflux pump inhibitor	*Campylobacter jejuni*	[95]
	Kaempferol	*Alpinia calcarata*	Efflux pump inhibitor	*Methicillin-resistant S. aureus*	[96]
	Caffeic acid	*Coffea arabica, Euphorbia hirta (L.)*	Cell membrane disruption	*Methicillin-resistant S. aureus, Listeria monocytogenes, E. coli and Salomonella typhimurium, P. aeruginosa*	[59, 97]
	Chlorogenic acid	*Ilex paraguariensis, Camellia sinensis*	Cell membrane disruption	*Methicillin-resistant S. aureus, Listeria monocytogenes, E. coli and Salomonella typhimurium*	[59, 98]
	Ferulic acid	*Ferula foetida*	Cell membrane disruption	*Methicillin-resistant S. aureus, Listeria monocytogenes, E. coli and Salomonella typhimurium*	[59, 99]
	Gallic acid	*Terminalia chebula*	Efflux pump inhibitor	*Mannheimia haemolytica, Pasteurella multocida, E. coli, Pseudomonas spp.*	[99, 100]
Terpenes	Eugenol	*Anethum graveolens*	Cell membrane disruption	*Methicillin-resistant S. aureus*	[101]
	Linalool	*Cinnamomum verum*	Cell membrane disruption	*K. pneumoniae carbapenemase*	[102]
	Carvacrol	*Thymus capitatus*	Cell membrane disruption	*E. coli, E. aerogenes, S. aureus, P. aeruginosa*	[103]
	Geraniol	*Cymbopogon citratus*	Efflux pump inhibitor	*Enterobacter aerogenes*	[104]
	Cinnamaldehyde	*Cinnamomum verum*	Cell membrane disruption	*E. coli, S. aureus*	[105]
	Farnesol	*Cymbopogon nardus*	Cell membrane disruption	*S. aureus*	[106]
	Thymol	*Thymus vulgaris*	Cell membrane disruption	*S. aureus, E. coli*	[62]

Phytochemical class	Phytochemicals	Source	Mode of action	Reported pathogens	Reference
Coumarins	Galbanic acid	*Ferula szowitsiana*	Efflux pump inhibitor	*S. aureus*	[107]
	Aegelinol	*Ferulago campestris*	DNA gyrase inhibitor	*E. aerogenes, S. enterica serovar Typhi, Enterobacter cloacae, S. aureus*	[108]
	Asphodelin A	*Asphodelus microcarpus*	DNA gyrase inhibitor	*S. aureus, E. coli Pseudomonas aeruginosa*	[109]
	Osthole	*Prangos hulusii*	DNA gyrase inhibitor	*B. subtilis, S. aureus, K. pneumoniae*	[110]

Table 1.
List of antioxidant phytochemicals summarized for their antimicrobial properties, chemical class, source, and major mode of action against pathogenic bacterial strains.

silver nanoparticles may improve RONS production and, as a result, can improve the treatment's efficacy. These unique therapeutic techniques have the potential to improve the antibacterial activity of some repurposable medicines. Antimicrobials that have been clinically approved to treat common infections could even be employed in combination therapy against novel multidrug-resistant bacteria. Natural compounds can not only provide novel antimicrobial treatment options but can also lower the cost of antibiotics treatments in combination, reduce the dose of antibiotics, and therefore can slow the resistance development. Phytochemicals causing oxidative stress with multitarget mode of action further warrant less chance of developing resistance in pathogens. However, the majority of this information is currently based on *in vitro* trials, and further study is needed to show that these innovative medicines are effective against harmful bacteria *in-vivo*.

8. Conclusion and future perspectives

Medicinal and aromatic plants are an appealing source for novel therapeutics in the era of antibiotic-resistant "superbugs." Plants frequently produce phytochemicals as pathogen-defeating compounds. Many phytochemicals derived from various plants, showed promising antibacterial, antifungal, and antiviral action against a variety of human diseases to date. Phytochemicals offer a lot of potentials when it comes to managing and treating microbial infections and wounds. Antibacterial mechanism of action of several phytochemicals is well-known, and knowledge of these bioactive substances has exploded in recent years. In general, phytochemicals disrupt the bacterial membrane, reduce certain virulence factors such as enzymes and toxins, and prevent the formation of bacterial biofilms, etc. Antimicrobial, antioxidant, and wound-healing phytochemicals promote blood coagulation, infection prevention, and wound healing. Phytochemicals, with dual potential of antioxidants and antimicrobial activity, in alone or combined with antibiotics can not only boost the human immune response to fight infection but can also present newer treatment strategies to combat drug-resistant microbes. Natural compounds such as curcumin, and carotenoids are antioxidants themselves but are also known to modulate Nrf2. Thus, identification, evaluation, and formulation of such natural antimicrobials with dual oxidative stress and antioxidant actions can play an important role to cure infectious diseases while at the same time can repair ROS-induced stress damage and inflammation in the host. It is critical to research and evaluate all accessible solutions that can combat infections, and antimicrobial resistance, and simultaneously improve human lives. Phytochemicals are not only less expensive and more accessible, but they are also safer, less toxic, and have wider acceptance than allopathic pharmaceuticals. However, before suggesting phytochemicals for medicinal purposes, standardization, safety, and scientific evaluation are required.

Acknowledgements

The authors are thankful for the support and encouragement of the Director, CSIR-Central Institute of Medicinal and Aromatic Plants in Lucknow, India, and Dr. Bhavana Gangwar, also acknowledge the Indian Council of Medical Research (ICMR) in New Delhi, India.

Author details

Bhavana Gangwar[1*], Santosh Kumar[2] and Mahendra P. Darokar[1]

1 Bioprospection and Product Development Division, CSIR-Central Institute of Medicinal and Aromatic Plants, Lucknow, India

2 Metabolic Engineering and Fermentation Science Group, Department of Food Science, University of Wisconsin-Madison, Madison, WI, USA

*Address all correspondence to: bhavanagangwar1@gmail.com

References

[1] Ventola CL. The antibiotic resistance crisis: Part 1: Causes and threats. Pharmacy and Therapeutics. 2015;**40**(4):277-283

[2] Centers for Disease Control and Prevention. Antibiotic Resistance Threats in the United States, 2019. Atlanta, GA: U.S. Department of Health and Human Services, CDC; 2019

[3] Kadri SS. Key takeaways from the U.S. CDC's 2019 antibiotic resistance threats report for frontline providers. Critical Care Medicine. 2020;**48**(7):939-945

[4] Weidinger A, Kozlov AV. Biological activities of reactive oxygen and nitrogen species: Oxidative stress versus signal transduction. Biomolecules. 2015;**5**(2):472-484

[5] Pohanka M. Role of oxidative stress in infectious diseases: A review. Folia Microbiologica. 2013;**58**(6):503-513

[6] Nathan C, Shiloh MU. Reactive oxygen and nitrogen intermediates in the relationship between mammalian hosts and microbial pathogens. Proceedings of the National Academy of Sciences U S A. 2000;**97**(16):8841-8848

[7] Piacenza L, Trujillo M, Radi R. Reactive species and pathogen antioxidant networks during phagocytosis. The Journal of Experimental Medicine. 2019;**216**(3):501-516

[8] Mourenza Á, Gil JA, Mateos LM, Letek M. Oxidative stress-generating antimicrobials, a novel strategy to overcome antibacterial resistance. Antioxidants. 2020;**9**(5):361

[9] Jensen PØ, Givskov M, Bjarnsholt T, Moser C. The immune system vs. Pseudomonas aeruginosa biofilms. FEMS Immunology and Medical Microbiology. 2010;**59**(3):292-305

[10] Müller A, Eller J, Albrecht F, Prochnow P, Kuhlmann K, Bandow JE, et al. Allicin induces thiol stress in bacteria through S-Allylmercapto modification of protein cysteines. The Journal of Biological Chemistry. 2016;**291**(22):11477-11490

[11] Alves NF, de Queiroz TM, de Almeida TR, Magnani M, de Andrade Braga V. Acute treatment with lauric acid reduces blood pressure and oxidative stress in spontaneously hypertensive rats. Basic & Clinical Pharmacology & Toxicology. 2017;**120**(4):348-353

[12] Adeyemi OS, Obeme-Imom JI, Akpor BO, Rotimi D, Batiha GE, Owolabi A. Altered redox status, DNA damage and modulation of L-tryptophan metabolism contribute to antimicrobial action of curcumin. Heliyon. 2020;**6**(3): e03495

[13] Upadhyay R, Chaurasia JK, Tiwari KN, Singh K. Antioxidant property of aerial parts and root of Phyllanthus fraternus Webster, an important medicinal plant. The Scientific World Journal. 2014;**2014**:692392

[14] Gonelimali FD, Lin J, Miao W, et al. Antimicrobial properties and mechanism of action of some plant extracts against food pathogens and spoilage microorganisms. Frontiers in Microbiology. 2018;**9**:1639

[15] Khare T, Anand U, Dey A, et al. Exploring phytochemicals for combating antibiotic resistance in microbial pathogens. Frontiers in Pharmacology. 2021;**12**:720726

[16] Barbieri R, Coppo E, Marchese A, Daglia M, Sobarzo-Sánchez E, Nabavi SF, et al. Phytochemicals for human disease: An update on plant-derived compounds antibacterial activity. Microbiological Research. 2017;**196**:44-68

[17] Pizzino G, Irrera N, Cucinotta M, et al. Oxidative stress: Harms and benefits for human health. Oxidative Medicine and Cellular Longevity. 2017;**2017**:8416763

[18] Stahl W, Sies H. Antioxidant defense: Vitamins E and C and carotenoids. Diabetes. 1997;**46**(Suppl 2):S14-S18

[19] Tan BL, Norhaizan ME, Liew WP, Sulaiman RH. Antioxidant and oxidative stress: A mutual interplay in age-related diseases. Frontiers in Pharmacology. 2018;**9**:1162

[20] Poljsak B, Šuput D, Milisav I. Achieving the balance between ROS and antioxidants: When to use the synthetic antioxidants. Oxidative Medicine and Cellular Longevity. 2013;**2013**:956792

[21] Rugemalira E, Roine I, Kuligowski J, Sánchez-Illana Á, Piñeiro-Ramos JD, Andersson S, et al. Protein oxidation biomarkers and myeloperoxidase activation in cerebrospinal fluid in childhood bacterial meningitis. Antioxidants. 2019;**8**(10):441

[22] Winterbourn CC, Kettle AJ, Hampton MB. Reactive oxygen species and neutrophil function. Annual Review of Biochemistry. 2016;**85**:765-792

[23] Nguyen GT, Green ER, Mecsas J. Neutrophils to the ROScue: Mechanisms of NADPH oxidase activation and bacterial resistance. Frontiers in Cellular and Infection Microbiology. 2017;**7**:373

[24] Heyworth PG, Cross AR, Curnutte JT. Chronic granulomatous disease. Current Opinion in Immunology. 2003;**15**(5):578-584

[25] Winterbourn CC. Toxicity of iron and hydrogen peroxide: The Fenton reaction. Toxicology Letters. 1995;**82-83**:964

[26] Fang YZ, Yang S, Wu G. Free radical homeostasis. Sheng Li Ke Xue Jin Zhan. 2004;**35**(3):199-204

[27] Winterbourn CC, Hampton MB. Thiol chemistry and specificity in redox signaling. Free Radical Biology & Medicine. 2008;**45**(5):549-561

[28] Winterbourn CC, Kettle AJ. Redox reactions and microbial killing in the neutrophil phagosome. Antioxidants & Redox Signaling. 2013;**18**(6):642-660

[29] Slauch JM. How does the oxidative burst of macrophages kill bacteria? Still an open question. Molecular Microbiology. 2011 May;**80**(3):580-583

[30] Teramoto H, Inui M, Yukawa H. OxyR acts as a transcriptional repressor of hydrogen peroxide-inducible antioxidant genes in Corynebacterium glutamicum R. The FEBS Journal. 2013 Jul;**280**(14):3298-3312

[31] Li N, Sioutas C, Cho A, Schmitz D, Misra C, Sempf J, et al. Ultrafine particulate pollutants induce oxidative stress and mitochondrial damage. Environmental Health Perspectives. 2003;**111**(4):455-460

[32] Comtois SL, Gidley MD, Kelly DJ. Role of the thioredoxin system and the thiol-peroxidases Tpx and Bcp in mediating resistance to oxidative and nitrosative stress in Helicobacter pylori. Microbiology. 2003;**149**(Pt 1):121-129

[33] Saleh M, Bartual SG, Abdullah MR, et al. Molecular

architecture of Streptococcus pneumoniae surface thioredoxin-fold lipoproteins crucial for extracellular oxidative stress resistance and maintenance of virulence. EMBO Molecular Medicine. 2013;**5**(12):1852-1870

[34] Achard P, Gusti A, Cheminant S, Alioua M, Dhondt S, Coppens F, et al. Gibberellin signaling controls cell proliferation rate in Arabidopsis. Current Biology. 2009;**19**(14):1188-1193

[35] Majnooni MB, Fakhri S, Shokoohinia Y, Kiyani N, Stage K, Mohammadi P, et al. Phytochemicals: Potential therapeutic interventions against coronavirus-associated lung injury. Frontiers in Pharmacology. 2020;**11**:588467

[36] Le-Vo HN, Tran PT, Le L, Matsumoto Y, Motooka D, Nakamura S, et al. Complex Class 1 Integron in a clinical *Escherichia coli* strain from Vietnam Carrying Both mcr-1 and blaNDM-1. Frontiers in Microbiology. 2019;**10**:2472

[37] Rex DK, Repici A, Gross SA, Hassan C, Ponugoti PL, Garcia JR, et al. High-definition colonoscopy versus Endocuff versus EndoRings versus full-spectrum endoscopy for adenoma detection at colonoscopy: A multicenter randomized trial. Gastrointestinal Endoscopy. 2018;**88**(2):335-344

[38] Xu M, Wu P, Shen F, Ji J, Rakesh KP. Chalcone derivatives and their antibacterial activities: Current development. Bioorganic Chemistry. 2019;**91**:103133

[39] Wang S, Yao J, Zhou B, Yang J, Chaudry MT, Wang M, et al. Bacteriostatic effect of quercetin as an antibiotic alternative in vivo and its antibacterial mechanism in vitro. Journal of Food Protection. 2018;**81**(1):68-78

[40] Dias DA, Urban S, Roessner U. A historical overview of natural products in drug discovery. Metabolites. 2012;**2**(2):303-336

[41] Vedam R, Pansare TA, Narula J. Contributions of ancient Indian knowledge to modern medicine and cardiology. Indian Heart Journal. 2021;**73**(5):531-534

[42] Kurhekar JV. Ancient and modern practices in phytomedicine. In: Preparation of Phytopharmaceuticals for the Management of Disorders. Academic Press, Elsevier; 2021. pp. 55-75

[43] Newman DJ, Cragg GM, Snader KM. The influence of natural products upon drug discovery. Natural Product Reports. 2000;**17**(3):215-234

[44] Harvey AL. Natural products in drug discovery. Drug Discovery Today. 2008;**13**(19-20):894-901

[45] Solecki RS. Shanidar IV, a Neanderthal flower burial in northern Iraq. Science. 1975;**190**:880-881

[46] Baldo F. Prediction of modes of action of components of traditional medicinal preparations. Physical Sciences Reviews. 2020;**5**(2):20180115

[47] Lanfranco G. Popular use of medicinal plants in the Maltese Islands. Ínsula. 1992;**1**:34-35

[48] Raskin I, Ripoll C. Can an apple a day keep the doctor away? Current Pharmaceutical Design. 2004;**10**(27):3419-3429

[49] Kashiwada Y, Wang HK, Nagao T, Kitanaka S, Yasuda I, Fujioka T, et al. Anti-AIDS agents. 30. Anti-HIV activity of oleanolic acid, pomolic acid, and structurally related triterpenoids. Journal of Natural Products. 1998;**61**(9):1090-1095

[50] Khan AH, Pessin JE. Insulin regulation of glucose uptake: A

complex interplay of intracellular signalling pathways. Diabetologia. 2002;**45**(11):1475-1483

[51] Rosenberg PH, Renkonen OV. Antimicrobial activity of bupivacaine and morphine. Anesthesiology. 1985;**62**(2):178-179

[52] Thawabteh A, Juma S, Bader M, Karaman D, Scrano L, Bufo SA, et al. The biological activity of natural alkaloids against herbivores, cancerous cells and pathogens. Toxins. 2019;**11**(11):656

[53] Mohammadzadeh N, Mehri S, Hosseinzadeh H. Berberis vulgaris and its constituent berberine as antidotes and protective agents against natural or chemical toxicities. Iranian Journal of Basic Medical Sciences. 2017;**20**(5):538-551

[54] Panche AN, Diwan AD, Chandra SR. Flavonoids: An overview. Journal of Nutritional Science. 2016;**5**:e47

[55] Cushnie TP, Lamb AJ. Antimicrobial activity of flavonoids. International Journal of Antimicrobial Agents. 2005;**26**(5):343-356

[56] Arima H, Ashida H, Danno G. Rutin-enhanced antibacterial activities of flavonoids against Bacillus cereus and Salmonella enteritidis. Bioscience, Biotechnology, and Biochemistry. 2002;**66**(5):1009-1014

[57] Kasprzak-Drozd K, Oniszczuk T, Stasiak M, Oniszczuk A. Beneficial effects of phenolic compounds on gut microbiota and metabolic syndrome. International Journal of Molecular Sciences. 2021;**22**(7):3715

[58] Miklasińska-Majdanik M, Kępa M, Wojtyczka RD, Idzik D, Wąsik TJ. Phenolic compounds diminish antibiotic resistance of *Staphylococcus aureus* clinical strains. International Journal of Environmental Research and Public Health. 2018;**15**(10):2321

[59] Zhao Y, Chen M, Zhao Z, Yu S. The antibiotic activity and mechanisms of sugarcane (*Saccharum officinarum L.*) bagasse extract against food-borne pathogens. Food Chemistry. 2015;**185**:112-118

[60] Kępa M, Miklasińska-Majdanik M, Wojtyczka RD, Idzik D, Korzeniowski K, Smoleń-Dzirba J, et al. Antimicrobial potential of caffeic acid against *Staphylococcus aureus* clinical strains. BioMed Research International. 2018;**2018**:7413504

[61] Yang W, Chen X, Li Y, Guo S, Wang Z, Yu X. Advances in pharmacological activities of terpenoids. Natural Product Communications. 2020;**15**(3)

[62] Trombetta D, Castelli F, Sarpietro MG, Venuti V, Cristani M, Daniele C, et al. Mechanisms of antibacterial action of three monoterpenes. Antimicrobial Agents and Chemotherapy. 2005;**49**(6):2474-2478

[63] Xu YF, Lian DW, Chen YQ, Cai YF, Zheng YF, Fan PL, et al. In vitro and in vivo antibacterial activities of patchouli alcohol, a naturally occurring tricyclic sesquiterpene, against *Helicobacter pylori* infection. Antimicrobial Agents and Chemotherapy. 2017;**61**(6):e00122

[64] Appalasamy S, Lo KY, Ch, ng SJ, Nornadia K, Othman AS, Chan LK. Antimicrobial activity of artemisinin and precursor derived from in vitro plantlets of Artemisia annua L. BioMed Research International. 2014;**2014**:215872

[65] Zhao LX, Wang ZW, Zheng CJ, et al. Synthesis, characterization and antibacterial activity of oleanolic acid

C-28 modification. Journal of Liaoning Normal University (Natural Science Edition). 2013;**36**:529-533

[66] Lü JM, Lin PH, Yao Q, Chen C. Chemical and molecular mechanisms of antioxidants: Experimental approaches and model systems. Journal of Cellular and Molecular Medicine. 2010;**14**(4):840-860

[67] Clifford T, Acton JP, Cocksedge SP, Davies K, Bailey SJ. The effect of dietary phytochemicals on nuclear factor erythroid 2-related factor 2 (Nrf2) activation: A systematic review of human intervention trials. Molecular Biology Reports. 2021;**48**(2):1745-1761

[68] Lee MT, Lin WC, Yu B, Lee TT. Antioxidant capacity of phytochemicals and their potential effects on oxidative status in animals - A review. Asian-Australasian Journal of Animal Sciences. 2017;**30**(3):299-308

[69] Patra AK. An overview of antimicrobial properties of different classes of phytochemicals. Dietary Phytochemicals and Microbes. 2012;**18**:1-32

[70] Rempe CS, Burris KP, Lenaghan SC, Stewart CNJ. The potential of systems biology to discover antibacterial mechanisms of plant phenolics. Frontiers in Microbiology. 2017;**8**:422

[71] Dong C, Zhou J, Wang P, et al. Topical therapeutic efficacy of ebselen against multidrug-resistant *Staphylococcus aureus* LT-1 targeting thioredoxin reductase. Frontiers in Microbiology. 2020;**10**:3016

[72] Zheng D, Huang C, Huang H, Zhao Y, Khan MRU, Zhao H, et al. Antibacterial mechanism of curcumin: A review. Chemistry & Biodiversity. 2020 Aug;**17**(8):e2000171

[73] Singh V, Pal A, Darokar MP. A polyphenolic flavonoid glabridin: Oxidative stress response in multidrug-resistant Staphylococcus aureus. Free Radical Biology & Medicine. 2015;**87**:48-57

[74] Gangwar B, Kumar S, Darokar MP. Glabridin averts biofilms formation in methicillin-resistant *Staphylococcus aureus* by modulation of the surfaceome. Frontiers in Microbiology. 2020;**11**:1779

[75] Kohanski MA, Dwyer DJ, Hayete B, Lawrence CA, Collins JJ. A common mechanism of cellular death induced by bactericidal antibiotics. Cell. 2007;**130**:797-810

[76] Maurya A, Dwivedi GR, Darokar MP, Srivastava SK. Antibacterial and synergy of clavine alkaloid lysergol and its derivatives against nalidixic acid-resistant Escherichia coli. Chemical Biology & Drug Design. 2013;**81**(4):484-490

[77] Sridevi D, Shankar C, Prakash P, Park JH, Thamaraiselvi K. Inhibitory effects of reserpine against efflux pump activity of antibiotic resistance bacteria. Chemical Biology Letters. 2017;**4**(2):69-72

[78] Boberek JM, Stach J, Good L. Genetic evidence for inhibition of bacterial division protein FtsZ by berberine. PLoS One. 2010;**5**(10):e13745

[79] Lamontagne Boulet M, Isabelle C, Guay I, et al. Tomatidine is a lead antibiotic molecule that targets *Staphylococcus aureus* ATP synthase subunit C. Antimicrobial Agents and Chemotherapy. 2018;**62**(6):e02197

[80] Xia LQ, Zhao BG, Ju YW, Dai L. Inhibitory effect of bis (piperidine) alkaloids on five environmental bacterial strains. Journal of Nanjing Medical University. 2001:81-84

[81] Yan Y, Li X, Zhang C, Lv L, Gao B, Li M. Research progress on antibacterial activities and mechanisms of natural alkaloids: A review. Antibiotics. 2021;**10**(3):318

[82] Bendaif H, Melhaoui A, Ramdani M, Elmsellem H, Douez C, El Ouadi Y. Antibacterial activity and virtual screening by molecular docking of lycorine from Pancratium foetidum Pom (Moroccan endemic Amaryllidaceae). Microbial Pathogenesis. 2018;**115**:138-145

[83] Rukachaisirikul T, Prabpai S, Champung P, Suksamrarn A. Chabamide, a novel piperine dimer from stems of Piper chaba. Planta Medica. 2002;**68**(9):853-855

[84] Salehi B, Venditti A, Sharifi-Rad M, et al. The therapeutic potential of apigenin. International Journal of Molecular Sciences. 2019;**20**(6):1305

[85] Osonga FJ, Akgul A, Miller RM, Eshun GB, Yazgan I, Akgul A, et al. Antimicrobial activity of a new class of phosphorylated and modified flavonoids. ACS Omega. 2019;**4**(7):12865-12871

[86] Plaper A, Golob M, Hafner I, Oblak M, Šolmajer T, Jerala R. Characterization of quercetin binding site on DNA gyrase. Biochemical and Biophysical Research Communications. 2003;**306**:530-536

[87] Ouyang J, Sun F, Feng W, Sun Y, Qiu X, Xiong L, et al. Quercetin is an effective inhibitor of quorum sensing, biofilm formation and virulence factors in Pseudomonas aeruginosa. Journal of Applied Microbiology. 2016;**120**(4):966-974

[88] Wu T, He M, Zang X, Zhou Y, Qiu T, Pan S, et al. A structure-activity relationship study of flavonoids as inhibitors of *E. coli* by membrane interaction effect. Biochimica et Biophysica Acta. 2013;**1828**(11):2751-2756

[89] Xu H, Ziegelin G, Schröder W, et al. Flavones inhibit the hexameric replicative helicase RepA. Nucleic Acids Research. 2001;**29**(24):5058-5066

[90] Cushnie TP, Lamb AJ. Assessment of the antibacterial activity of galangin against 4-quinolone resistant strains of Staphylococcus aureus. Phytomedicine. 2006;**13**(3):187-191

[91] Ming J, Zhuoneng L, Guangxun Z. Protective role of flavonoid baicalin from Scutellaria baicalensis in periodontal disease pathogenesis: A literature review. Complementary Therapies in Medicine. 2018;**38**:11-18

[92] Ozma MA, Khodadadi E, Pakdel F, Kamounah FS, Yousefi M, Yousefi B, et al. Baicalin, a natural antimicrobial and anti-biofilm agent. Journal of Herbal Medicine. 2021;**27**:100432

[93] Jeong KW, Lee JY, Kang DI, Lee JU, Shin SY, Kim Y. Screening of flavonoids as candidate antibiotics against Enterococcus faecalis. Journal of Natural Products. 2009;**72**(4):719-724

[94] Chan BC, Ip M, Lau CB, Lui SL, Jolivalt C, Ganem-Elbaz C, et al. Synergistic effects of baicalein with ciprofloxacin against NorA over-expressed methicillin-resistant *Staphylococcus aureus* (MRSA) and inhibition of MRSA pyruvate kinase. Journal of Ethnopharmacology. 2011;**137**(1):767-773

[95] Klančnik A, Šikić Pogačar M, Trošt K, Tušek Žnidarič M, Mozetič Vodopivec B, Smole MS. Anti-Campylobacter activity of resveratrol and an extract from waste Pinot noir

grape skins and seeds, and resistance of Camp. jejuni planktonic and biofilm cells, mediated via the CmeABC efflux pump. Journal of Applied Microbiology. 2017;**122**(1):65-77

[96] Randhawa H, Hundal K, Ahirrao P, Jachak S, Nandanwar H. Efflux pump inhibitory activity of flavonoids isolated from Alpinia calcarata against methicillin-resistant Staphylococcus aureus. Biologia. 2016;**71**(5):484-493

[97] Perumal S, Mahmud R, Ismail S. Mechanism of action of isolated caffeic acid and Epicatechin 3-gallate from Euphorbia hirta against *Pseudomonas aeruginosa*. Pharmacognosy Magazine. 2017;**13**(2):S311-S315

[98] Lou Z, Wang H, Zhu S, Ma C, Wang Z. Antibacterial activity and mechanism of action of chlorogenic acid. Journal of Food Science. 2011;**76**(6):M398

[99] Borges A, Ferreira C, Saavedra MJ, Simões M. Antibacterial activity and mode of action of ferulic and gallic acids against pathogenic bacteria. Microbial Drug Resistance. 2013;**19**(4):256-265

[100] Rajamanickam K, Yang J, Sakharkar MK. Gallic acid potentiates the antimicrobial activity of tulathromycin against two key bovine respiratory disease (BRD) causing-pathogens. Frontiers in Pharmacology. 2019;**9**:1486

[101] Yadav MK, Chae SW, Im GJ, Chung JW, Song JJ. Eugenol: A phyto-compound effective against methicillin-resistant and methicillin-sensitive Staphylococcus aureus clinical strain biofilms. PLoS One. 2015;**10**(3):e0119564

[102] Yang SK, Yusoff K, Ajat M, Thomas W, Abushelaibi A, Akseer R, et al. Disruption of KPC-producing Klebsiella pneumoniae membrane via induction of oxidative stress by cinnamon bark (Cinnamomum verum J. Presl) essential oil. PLoS One. 2019;**14**(4):e0214326

[103] Althunibat OY, Haitham Q, Sati YA, Muayad A, Khaled K, Ibrahim SM, et al. Effect of thymol and carvacrol, the major components of thymus capitatus on the growth of *Pseudomonas aeruginosa*. Journal of Pure and Applied Microbiology. 2016;**10**(1):367-374

[104] Lorenzi V, Muselli A, Bernardini AF, Berti L, Pagès JM, Amaral L, et al. Geraniol restores antibiotic activities against multidrug-resistant isolates from gram-negative species. Antimicrobial Agents and Chemotherapy. 2009;**53**(5):2209

[105] Zhang Y, Liu X, Wang Y, Jiang P, Quek SY. Antibacterial activity and mechanism of cinnamon essential oil against Escherichia coli and *Staphylococcus aureus*. Food Control. 2016;**59**:282-289

[106] Togashi N, Inoue Y, Hamashima H, Takano A. Effects of two terpene alcohols on the antibacterial activity and the mode of action of farnesol against *Staphylococcus aureus*. Molecules. 2008;**13**(12):3069-3076

[107] Bazzaz Fazly BS, Iranshahi M, Naderinasab M, Hajian S, Sabeti Z, Masumi E. Evaluation of the effects of galbanic acid from Ferula szowitsiana and conferol from F. badrakema, as modulators of multi-drug resistance in clinical isolates of Escherichia coli and Staphylococcus aureus. Research in Pharmaceutical Sciences. 2010;**5**(1):21-28

[108] Basile A, Sorbo S, Spadaro V, Bruno M, Maggio A, Faraone N, et al. Antimicrobial and antioxidant activities of coumarins from the roots of Ferulago campestris (Apiaceae). Molecules. 2009;**14**(3):939-952

[109] El-Seedi HR. Antimicrobial arylcoumarins from Asphodelus microcarpus. Journal of Natural Products. 2007;**70**(1):118-120

[110] Tan N, Yazıcı-Tütüniş S, Bilgin M, Tan E, Miski M. Antibacterial activities of pyrenylated Coumarins from the Roots of *Prangos hulusii*. Molecules. 2017;**22**(7):1098

Chapter 10

Involvement of Antioxidant in the Prevention of Cellular Damage

Olalekan Bukunmi Ogunro, Aderonke Elizabeth Fakayode and Gaber El-Saber Batiha

Abstract

Oxidative stress occurs when the body's enzymatic or non-enzymatic antioxidants are outweighed by endogenous or exogenous free radicals. Oxidative radicals, reactive oxygen species, and other biomolecule-damaging free radicals can be generated during normal cellular metabolism and react with proteins, lipids, and DNA. In the domains of biology and medicine, free radicals have become increasingly important. They can accumulate in a variety of ways, both endogenously and exogenously. Mitochondria are the primary source of cell-level endogenous reactive oxygen species. In several chronic and degenerative disorders, this results in tissue destruction. In addition to being produced endogenously, antioxidants can also be delivered exogenously to the biological system, most frequently through nutrition. Antioxidants are generally used to counteract the effects of free radicals produced by metabolic processes. In this chapter, the crucial function of reactive oxygen species in human health, as well as exploring the functioning of antioxidative defense systems in reducing toxicity caused by excess reactive oxygen species were discussed.

Keywords: cellular damage, free radicals, human diseases, cell signaling, antioxidant milieu

1. Introduction

Defending cells against the harm produced by free radicals is the goal of antioxidants. Taking antioxidants may help to counteract some of the damage that free radicals can inflict. Carotenoids and other nutrients like beta carotene and lycopene have been shown to be effective antioxidants [1]. Free radicals can be generated during oxidation reactions, which can set off a cascade of events that damage cells. By eliminating free radical intermediates, antioxidants put an end to these chain events and block further oxidation reactions. They participate in processes that maintain cell health and repair DNA. Because of this, oxidizing substances like thiols or ascorbic acid frequently serve as antioxidants. Plants and animals utilize a wide variety of antioxidants, such as glutathione and vitamins C and E, as well as enzymes such as catalase, superoxide dismutase, glutathione peroxidase to mitigate the deleterious effects of oxidation reactions [2]. If a cell's antioxidant levels are low or its antioxidant enzymes are blocked, oxidative stress can cause the cell to become damaged [3, 4].

A wide variety of dietary supplements contain antioxidants which help people stay healthy while also reducing their risk of developing ailments like cancer and heart disease. In later, more extensive clinical tests, researchers were unable to demonstrate any benefit to taking antioxidant supplements; instead, they discovered that taking an excessive amount may be harmful [5]. Natural antioxidants have a variety of applications in industry, in addition to their usage in medicine. Some of these applications include acting as preservatives in food and cosmetics [6, 7].

2. Oxidants and free radicals

Any molecular species that has an unpaired electron in an atomic orbital and is capable of independent existence is referred to as a free radical. When an electron is missing from a pair, it causes the resulting species to be extremely reactive. Free radicals are capable of a diverse set of reactions, the most common of which are electron transfer and addition processes that lead to the creation of covalent bonds [8, 9]. Reducing free radicals are those that give up an electron to an acceptor, while oxidizing free radicals are those that take in electrons (accepting an electron from a donor). There is a thermodynamic hierarchy, often known as a pecking order, for the many types of electron transfer reactions. This is because radicals can have a wide variety of reactivities [1, 9].

2.1 Generation of free radicals and oxidants

Non-radicals can be converted into radicals through a variety of methods, including the addition of a single electron to the molecule. A covalent bond (C–H, C–O or C–C) can be broken via homolytic fission, in which one electron from the bonding pair remains on each atom. While disulfide links can readily be broken, the O–O bond in H_2O_2 can be broken by exposing it to UV light, resulting in the formation of $^{\bullet}OH$, these covalent bonds are extremely difficult to dissociate. Exogenous and endogenous sources of free radicals exist [10, 11]. Different cell organelles, such as mitochondria, peroxisomes, and endoplasmic reticulum, as well as various enzyme activity, fatty acid metabolism, and phagocytic cells, are examples of endogenous sources. High temperatures and environmental pollutants, such as those produced by cooking (smoked meat or used cooking oil), H_2O_2, N_2O_2, deoxyosones, and ketamine, are examples of exogenous sources of radiation. Other exogenous sources include X-ray and beta-ray light, ultraviolet A light in the presence of a sensitizer and chemical reagents such as these (aromatic hydrocarbons, pesticides, polychlorinated biphenyls, dioxins, and many others), microbial infections, drugs, and their metabolites [9, 11]. To combat bacteria and other invaders, activated immune cells (eosinophils, neutrophils, etc.) produce endogenous free radicals, as does the mitochondrial respiratory chain, enzymatic activity (xanthin oxidase, NADPH oxidase, lipo-oxygenase, NO synthase, etc.), and various pathological conditions and diseases [12, 13]. Air and water pollution, cigarette smoke, heavy metals or transition and other medications and chemicals, radiation and extreme temperatures produce exogenous free radicals.

3. Oxidative damage to cellular molecules

3.1 Oxidative damage to protein

Free radicals, amino acid modification, cross-linkage formation due to lipid peroxidation, and protein fragmentation are all methods by which proteins can be damaged. Methionine, cysteine, arginine, and histidine are the most susceptible to oxidation in proteins. Proteins that have already been damaged by free radicals are more vulnerable to enzyme proteolysis. The oxidation of protein products can influence enzymes, receptors, and the transport of molecules across membranes [13].

Since oxidatively damaged protein products contain highly reactive groups, membrane damage and other cellular activities may be impaired because of their existence. Peroxyl radicals, a type of free radical, are hypothesized to be responsible for protein oxidation. Carbonyls and other amino acid modifications can be created because of ROS damaging proteins and causing carbonyl and other amino acid alterations, such as the production of methionine sulfoxide and protein peroxide. From signaling pathways to enzyme activity to heat stability to proteolysis susceptibility, many elements of protein oxidation are affected [4, 9].

3.2 Lipid peroxidation

In a variety of physiological and pathological processes, Including aging, arterial hardening, inflammation, and cancer development and progression, oxidative stress play an important role to increase biochemical lesions by reacting with other biomolecules [14]. Cell membrane-bound polysaturated fatty acids are subjected to lipid peroxidation, which progresses via radical chain reaction. ROS is hypothesized to be triggered by hydroxyl radicals, which remove hydrogen atoms, resulting in the formation of lipid radicals and diene conjugates. In addition, it generates a peroxyl radical when oxygen is added; this extremely reactive radical then attacks a different fatty acid, resulting in lipid hydroperoxide (LOOH) and a brand-new radical. As a result, lipid peroxidation grows. Several chemicals are generated because of lipid peroxidation, including alkanes, malonaldehyde, and isoprotanes. Researchers have shown these chemicals to be biomarkers for lipid peroxidation in a variety of conditions including diabetes, ischemia reperfusion injury and neurodegenerative disorders [4, 15].

3.3 Oxidative damage to DNA

Oxidative DNA damage is an inevitable consequence of cellular metabolism. While guanine typically pairs with cytosine, the most common form of oxidative base damage, 8-oxo-7,8-dihydroguanine (8-oxoG), can lead to adenine mispairing through a conformational change. DNA and RNA can be damaged by oxidative stress, as numerous studies have demonstrated beyond reasonable doubt. Many diseases, including aging and cancer, have been linked to mutations in DNA. When free radicals or ultraviolet radiation cause oxidative damage to DNA, the levels of oxidative nucleotides including glycol, dTG, and 8-hydroxy-2-deoxyguanosine rise [16]. One of the many illnesses linked to oxidative damage is cancer, and mitochondrial DNA has been found to be particularly vulnerable. The use of 8-hydroxy-2-deoxyguanosine as a biomarker for oxidative stress is well adopted. Oxidatively stressed cells have high levels of this marker [17].

4. Biological activities of free radicals and oxidants

It is necessary for the maturation of cellular structures that both reactive oxygen species (ROS) and reactive nitrogen species (RNS) are present in low to moderate concentrations since they can operate as weapons for the host defense system. It is true that phagocytes (which include neutrophils, macrophages, and monocytes) create free radicals as a part of the body's immune system's fight against sickness. Reactive oxygen species (ROS) production by the immune system is clearly demonstrated in patients with granulomatous illness (ROS) [18]. The membrane-bound NADPH oxidase machinery is faulty in these individuals; therefore, they are unable to create the superoxide anion radical ($O_2^{\bullet}$). The effect is that people get sick and become infected with numerous diseases that endure for a long time.

Several cellular signaling systems benefit from the physiological functions ROS and RNS play in their operation, as well (ROS and RNS). Nonphagocytic NADPH oxidase isoforms are crucial for the control of intracellular signaling cascades in fibroblasts, endothelial cells, vascular smooth muscle cells, cardiac myocytes, and thyroid tissue. Blood flow, clotting, and cognitive function are all affected by nitric oxide (NO), an intercellular messenger [18, 19].

In addition to its role in nonspecific host defense, NO is essential for the eradication of intracellular infections and malignancies. A mitogenic reaction is one of the many good effects of free radicals. At low to moderate levels of intensity, ROS and RNS are required for human health.

5. Mechanism of cell signaling mediated by RNS/ROS

Oxidation can take place in any of these macromolecules: DNA, proteins, and lipids when ROS are present. Oxidative stress is primarily caused by reactive oxygen species (ROS) in cells. Signaling molecules such as ROS are critical to the proper operation of the body's physiological systems. In a physical sense, this is what is going on. Autophagy, apoptosis, necrosis, and other mechanisms that lead to cell death are activated when ROS levels are too high.

5.1 ROS induce autophagy

Lysosomes of the cell remove damaged organelles, protein aggregates, and foreign invaders via autophagy, a cellular breakdown process. Several human diseases, including as cancer, neurological disorders, infectious diseases, metabolic disorders, and the aging process, may be caused in part by problems with autophagy. Autophagy can be triggered in response to a variety of stresses, including starvation, ER stress, organelle breakdown, and pathogen infection. Activation of autophagy has been linked to reactive oxygen species (ROS). H_2O_2 will eventually cause oxidative stress because of its buildup in the cell. The autophagic process relies heavily on the autophagy gene ATG4. There is evidence to suggest that H_2O_2 oxidizes this gene specifically in the absence of food. If H_2O_2 builds up, the ATG4's activity can be oxidized. The lipidation of LC3/ATG8 is essential for the initiation of autophagy, which is facilitated by oxidized ATG4. For the buildup of LC3-PE on autophagosome membranes and the subsequent stimulation of autophagosome formation, ROS is necessary [20, 21].

Reactive oxygen species (ROS) have the potential to regulate autophagy via activating the mitogen-activated protein kinase [MAPK] family. JNK1c-Jun-N-terminal kinase (JNK), p38, and extracellular signal-regulated kinase are all members of this family (ERK). In the three-tier kinase cascade that activates the members of the MAPK family, MAPK kinase (MAPKK), MAPK kinase (MAPKK), and MAPK all participate. When JNK is activated for an extended period, the cell's production of reactive oxygen species (ROS) increases significantly, increasing the risk of DNA damage. The identification of cellular redox stress is the final step in the activation of the p53 pathway. Many autophagy-inducing genes can be activated by p53 as a transcription factor. As a result, JNK and Sestrin2 may be activated, resulting in the phosphorylation and activation of TSC2 and the resulting autophagy [22].

Additional signaling pathways that participate in ROS-mediated autophagy and contribute to the process include Akt/mTOR (mechanistic target of rapamycin), as well as AMPK The well-known kinase Akt/mTOR, which in turn oxidizes the phosphatase and tensin homolog, is controlled by reactive oxygen species (ROS) (PTEN). Inhibition of mTOR and activation of AMPK are required for autophagy activation, and these two processes are controlled by the VPS34 complex [23].

5.2 ROS trigger apoptosis

Death receptors and mitochondrial pathways initiate cell apoptosis in response to both external and internal stimuli. Because an increase in oxidative stress disturbs the homeostatic equilibrium within cells and causes long-term oxidative changes to fat, protein, or DNA, ROS levels rise. TRAIL and nuclear factor kappa B (NF-kB) are activated by reactive oxygen species (ROS) and result in the death of cancer cells. Apoptosis can be induced by ROS-driven activation of JNK, a MAPK family member like JNK. Mitochondrial malfunction and apoptosis are becoming more obvious roles for JNK [24]. Several studies have shown that Shikonin, a naturally occurring naphthoquinone derivative, can kill cancer cells. Shikonin boosted ROS production and apoptosis, as well as the production of JNK and p38 in K562 cells, which were then treated. Programed cell death in cancer cells increases because of ROS/JNK activation [25]. The redox sensitive MAPK kinase and Apoptosis Signal Regulation Kinase 1 (ASK1) are positioned upstream of ROS/JNK. The antioxidant protein ASK1 is prevented from conducting its work by Grx and Trx1, which are antioxidant proteins. Components associated with the tumor necrosis factor receptor are recruited to the complex when ROS cause Trx1 to dissociate from the ASK1-Trx1. Activated ASK1 signals can activate AP-1-dependent proapoptotic genes and mitochondrial signaling. By altering the mitochondrial ASK1/ASK2/Trx2 complex, ROS can also lead to the release of cytochrome C. Increase ROS levels in the ER and stimulate mitochondria to do this. Antioxidant flavone can protect against myocardial ischemia/reperfusion injury, which can lead to apoptosis [26, 27].

5.3 Necrosis induced by ROS

In contrast to apoptosis, necrosis is a unique form of cell death. The receptor-interacting serine/threonine 3-like (RIP3) protein kinase has the potential to destroy cells because it is highly expressed in so many different cell lines. The RIP1 and RIP3 serine/threonine kinases both regulate necrosis in a similar manner. To activate the transmission of the pro-necrotic signal, RIP1 and RIP3 must be phosphorylated to

form necrosome, an amyloid-like complex. Depletion of RIP3 in necrosis-inducing cells reduces ROS concentration, but RIP3 overexpression raises ROS levels.
The involvement of RIP3 in necrosis induction is performed via increasing ROS production associated to energy metabolism [28, 29]. When RIP1 and RIP3 are phosphorylated, the pronecrotic kinase activity is triggered, and ROS are generated. According to this study, the phosphorylation of pronecrotic complexes stabilizes their interactions. A link between STAT3 and the mitochondrial electron transport chain complex I component GRIM-19 governs enhanced ROS production from RIP1 phosphorylation-dependent activation of the mitochondrial electron transport chain. In the mitochondria, STAT3 and GRIM-19 accumulate and increase ROS production and necroptosis, the process by which cells die, because of this interplay [30]. They all work together to increase energy consumption and mitochondrial ROS generation by enhancing the interaction between RIP3 and the enzymes glutamate dehydrogenase 1 [GDH], glutamate ligase (GLUL), and PYGL. During necrosis induction, there may be an interaction between the TNF receptor and the necrosis-inducing ROS generated by the NADPH oxidase NOX enzyme complex. As an important RIP3 downstream component of TNF-induced necrosis in ROS-induced necrosis, MLKL has been found to be an important player in the process. In the last phases of necrosis caused by TNF, MLKL also plays a role in ROS production and JNK activation [31].

6. Oxidative stress and human diseases

An organ or tissue is said to be under oxidative stress when the endogenous antioxidant defense system is overwhelmed by the production of highly reactive molecules like ROS, RNS and RSS, resulting in cellular damage and malfunction and a wide spectrum of illnesses. As a result of normal metabolic activities, the reactive species are created in low concentrations within the cells themselves. Radiation (X-rays and UV), pollution, cigarette smoke, bacteria, viruses, and drugs can also cause them, as can acute or chronic cellular stress (acute or chronic) [24]. They include free radicals and nonradical oxidants. Free radicals are unstable because of the presence of unpaired electrons in their outer electron orbit. Free radicals tend to neutralize themselves by reacting with other molecules and triggering their oxidation because they are so unstable and reactive. Thus, they have the potential to disrupt a wide spectrum of biological components, such as DNA, lipids, and proteins. Proteins are a common target for free radicals because of their critical role in cellular activity. Although free radicals have been shown to cause some protein modifications, such as protein unfolding or structural alteration, the majority are absolutely harmless. It is possible for protein inactivation and long-term cellular damage to be caused by irreversible protein alterations, even if reversible oxidative changes govern protein activity [1].

6.1 Oxidative stress in atherosclerosis

Atherosclerosis is a chronic inflammatory disease of the vascular system that is marked by chronic inflammation. There is a strong correlation between cardiovascular disease and atherosclerosis in most developed countries (CVD). The endothelium is injured because of inflammation and oxidative stress, resulting in arterial lesions and plaque deposition [32]. It is easier for plaque, which is mostly composed of blood cells and foam cells as well as lipids and proteins to impede the vascular system and prevent blood flow. Infarctions and strokes resulting from coronary artery disease characterize cardiovascular disease (CVD). Diabetes, high blood pressure, smoking, cholesterol difficulties, obesity, and other metabolic illnesses are linked to endothelial degradation. In the early phases of atherosclerosis, oxidative stress has a negative impact on endothelial function. Endothelial function, inflammation, bleeding, and oxidative damages are all influenced by the endocrine system (RAS) [33, 34]. As a result of activating NADPH oxidase in the cardiovascular system, reactive oxygen species (ROS) are produced that damage the endothelium, resulting in endothelial dysfunction (ROS).

6.2 Oxidative stress in hypertension

The most common cause of cardiovascular disease and death around the world is high blood pressure. About 90% of instances of hypertension are categorized as essential hypertension, when the exact reason is unknown. Hypertensive stimuli, such as salt and the hyperactive RAS and OS systems as well as endogenous hormones such as Ang II and aldosterone, produce protein modification that results in a rise in blood pressure. Neoantigens are proteins that have been altered so that they are no longer identified by activated T cells as being their own. Macrophages in the blood and kidneys are stimulated to release proinflammatory cytokines by T cell-derived signals. Activated T cells in the vasculature enhance renal salt and water retention, as well as renal vasoconstriction and remodeling. Chronic inflammation can lead to high blood pressure, which is a risk factor for OS. In the presence of Ang II-induced hypertension, T cells show substantial amounts of p47phox, p22phox, and NOX_2 oxidase components. To put it another way, the transfer of faulty T cells results in arterial hypertension and a decreased generation of oxygen. Angiotensin II (Ang II) is one of the most major ROS generators, while NADPH oxidase is one of the most prominent ROS producers [35, 36]. The production of Ang II reaches its peak under hypertensive conditions. In addition, increased angiotensin II levels can promote necrosis and apoptosis in renal tissue during the period of reperfusion. Ang II inhibits the SR-BI HDL receptor in proximal tubular cells. Statins were intended to inhibit HMG-CoA reductase to lower cholesterol production. However, these medications have anti-inflammatory properties as well as the potential to reduce systolic blood pressure in people with high cholesterol as part of their pleiotropic effects. Patients with elevated blood pressure feel the effects more intensely [37].

6.3 Oxidative stress in diabetes mellitus

The body's ability to neutralize free radicals and produce antioxidants is out of whack, resulting in diabetes mellitus (DM). Diabetes can be triggered by changes in blood glucose levels. OS has a significant impact on the emergence of DM problems. A high blood sugar level can have a significant impact on a person's overall health [38]. As a result, chronic hyperglycemia has a lower OS than any other kind of glucose oscillation. Long-term and severe chronic hyperglycemia, as well as frequent blood glucose fluctuations, are hallmarks of many glycemic disorders. Hyperglycemia triggers ROS production in the body. Even if the cells of persons with type 2 diabetes are still functioning and intact, ROS produces OS because of the existence of ROS. Insulin production is reduced as a result. Diabetes mellitus has been linked to an increase in the radical O_2 - in both animal and vitro investigations. Many mechanisms, such as enzymatic, nonenzymatic, and mitochondrial processes, exist in DM for the generation of oxidative stress. There are numerous reasons for the rise in OS in DM. The most major oxidizing activity, glucose autooxidation, generates free radicals [38, 39]. Reduced antioxidant defenses (lower levels of cellular antioxidants and decreased enzyme activity against free radicals) and unbalanced reduction/oxidation are also contributing factors. High blood glucose levels activate numerous pathways when O_2 - is generated, for the same reasons the hexosamine route is more active and the protein kinase C isoform is activated in DM. When studying DM in mitochondria, researchers look at how much energy is produced, ROS are produced, signals are transmitted, and cells die. The processes of mitochondrial fusion and fission are essential for the preservation of homeostasis. The expansion of the mitochondrial network via mitochondrial fusion appears to be beneficial. Excessive mitochondrial fission, which results in a buildup of mitochondrial fragments and a shortened electron transport chain, can aggravate cellular mitochondrial ROS generation [40].

6.4 Oxidative stress in neurodegenerative diseases

Alzheimer's, Parkinson's, and depression are all linked to OS. The emergence of neurological diseases like Alzheimer's and Parkinson's, both of which are intimately linked to aging, is a key risk factor for OS. Oxidative stress and mitochondrial dysfunction are two of OS's long-term side effects. The hippocampus of Alzheimer's disease animal models shows decreased activation of mitochondrial complex IV. As well as causing mitochondrial oxidative damage, increased OS also generates harmful byproducts for the brain [41, 42]. Alzheimer's disease neurodegeneration is linked to the production of a potentially hazardous peptide known as -amyloid by ROS. Neocortical neurons produce more H_2O_2 when -amyloid is present. Activation of NADPH by microglia cells in Parkinson's disease mice is also linked to the progression of dopaminergic neurodegeneration. Multiple sclerosis (MS) and depressive and autoimmune illnesses are all connected to OS. Multiple sclerosis patients have lower GPx enzyme activity and higher levels of oxidative damage to DNA (8-OHdG). Patients with unipolar depression have been shown to have low levels of SOD, ascorbic acid, and MDA [43].

6.5 Oxidative stress in cancer

Cancerous cells can proliferate more quickly when ROS is present. OH, the major ROS that damages mitochondrial and nuclear DNA, hydrolyzes bases to form 8-OHdG and 8-oxodG, two examples of hydrolyzed base products. Various enzymatic pathways can be used by cells to repair damaged DNA. However, mutations caused by base change or deletion can cause cancer if DNA damage is too severe to repair. Insufficient DNA repair is more likely when there are twice as many DNA oxidative damages [44]. As we age, the bodies' ability to repair oxidative damage and other forms of DNA damage decreases. Cytotoxicity and chromosomal disorders can result from DNA oxidation. Genetic mutations may be generated through free radical interactions with other biological components, in addition to DNA damage. The carcinogen LPO is a known carcinogen. In the presence of guanine bases, MDA may create adducts, which are toxic. It's yet unclear how OS-induced carcinogenesis affects the human body. The ability of OS to alter gene and protein expression that signals cell growth and proliferation has been demonstrated through new techniques [1, 24].

7. Classification of antioxidants

In order to protect cells from oxidative stress, antioxidant enzymes network together to create a protective barrier. Oxidative phosphorylation and other cellular processes generate H_2O_2, which is then reduced to water. The initial stage in this detoxification pathway is initiated by superoxide dismutase, and hydrogen peroxide is eliminated by catalases and other peroxidases [2, 45].

7.1 Enzymatic antioxidants

7.1.1 Superoxide dismutase

A set of enzymes known as superoxide dismutases (SODs) breaks down the superoxide anion to produce oxygen and hydrogen peroxide. SOD enzymes are found in almost all aerobic cells and fluids. It is possible to categorize superoxide dismutases into three main types based on their ability to bind iron or manganese as cofactors: the Cu/Zn, Fe, and Mn, and finally the Ni subtypes [46]. It has been discovered that SOD isozymes are present in several cell compartments in higher plants. Mn-SOD is found in both mitochondria and peroxisomes. CuZn-SOD was found in all four chloroplasts, peroxisomes, and apoplasts using fluorescent microscopy. Fe-SOD was also found in these organelles, albeit at lower concentrations [47].

Superoxide dismutase enzymes are found in all mammals and most chordates, including humans. SOD1 and SOD2 can be found in the cytoplasm, mitochondria, and extracellular space. In comparison, the other three are tetramers (four units) in structure (four subunits). SOD1 and SOD3 include copper and zinc, while SOD2 contains manganese in the reactive core [48, 49].

7.1.2 Catalase

Catalase is an antioxidant enzyme that also plays the role of a catalyst in the process of converting hydrogen peroxide into oxygen and water. It does this by nullifying the effect that the hydrogen peroxide that is present inside the cell would otherwise have. It is not possible to determine the exact quantity of catalase that is present in the cytoplasm due to the fact that the majority of it is destroyed whenever the tissue is handled. The interaction of reactive oxygen species and antioxidants can lead to an imbalance, which in turn can lead to oxidative stress. Oxidative stress is both a disease-causing and disease-aggravating factor, and it plays a role in the development of many different diseases [1, 50].

7.1.3 Glutathione systems

The glutathione system includes glutathione, glutathione reductase, glutathione peroxidases, and glutathione S-transferases. This system can be found in all living things, including bacteria, plants, and mammals. Hydroperoxides and hydrogen peroxide can be broken down by glutathione peroxidase, an enzyme found in the body. The enzyme Glutathione peroxidase possesses all four of the necessary selenium cofactors for this procedure. There are at least four unique isozymes of glutathione peroxidase reported in different species of animals. As far as scavenging hydrogen peroxide is concerned, glutathione peroxidase 1 is more common and more effective than glutathione peroxidase 4 when it comes to lipid hydroperoxides. S-transferase activity increases when lipid peroxides are present. These enzymes are particularly abundant in the liver, where they contribute to the metabolic process of detoxification [1].

7.2 Non-enzymatic antioxidants

7.2.1 Ascorbic acid

Both plants and animals contain ascorbic acid, a monosaccharide antioxidant. For this reason, the nutrient is categorized as a vitamin and must be ingested through food. Most other animals can produce this chemical on their own and do not need it in their diets because of that. Cells can continue to function effectively because glutathione is kept in a reduced state by protein disulfide isomerase and glutaredoxins. Ascorbic acid, for example, can neutralize ROS such as hydrogen peroxide (H_2O_2). In addition to being a direct antioxidant, ascorbic acid also provides a substrate for an enzyme called ascorbate peroxidase. Because of this function, plants are better equipped to deal with a wide range of stresses [51, 52].

7.2.2 Glutathione

In all aerobic living forms, the cysteine-containing peptide known as glutathione can be detected. It can be made in the cells of the body from the amino acids that make up its components, therefore getting it in one's food is not necessary for getting it. Because the thiol group in the cysteine that makes up glutathione is a reducing agent, glutathione could both oxidize and reduce itself in a reversible fashion, giving it antioxidant properties [53, 54]. Within cells, the enzyme glutathione reductase is responsible for maintaining the reduced form of glutathione. Glutathione, in this state, can reduce the levels of other metabolites and enzyme systems, as well as react directly with oxygen. A key antioxidant in cells, glutathione has an extremely high concentration and plays a critical role in maintaining the redox balance within cells. Mycothiol and trypanothione, two additional thiols can be substituted for glutathione in some organisms, such as actinomycetes and Kinetoplastids [55].

7.2.2.1 Thiols

The group of organic compounds with a sulfhydryl group includes thiols (-SH). They are made up of a carbon atom joined to a hydrogen atom and a sulfur atom. In the organism, extra electrons pass to thiols during the oxidation caused by ROS, resulting in the formation of disulphide bonds. These reversible bonds allow electrons to transfer back to thiols due to the oxidative balance. In enzymatic reactions, signal transduction, detoxification, transcription, regulation of enzymatic activation, cellular signaling mechanisms, and apoptosis reaction, thiol-disulphide homeostasis' antioxidant capacity plays a crucial role [55].

7.2.3 Tocopherols and tocotrienols (vitamin E)

Tocopherols and tocotrienols make form a set of eight different fat-soluble vitamins with antioxidant properties. These vitamins are closely connected to one another. The umbrella term "vitamin E" is used to refer to all these different vitamins. The most research has been done on beta-tocopherol because of its high bioavailability compared to the other tocopherols [56]. This demonstrates that the body can absorb beta-tocopherol and metabolize it more efficiently than it does the other forms. It has been hypothesized that the form of tocopherol known as gamma-tocopherol is the most effective lipid-soluble antioxidant, and that it shields membranes from oxidation by engaging in a chain reaction with lipid radicals that are generated during the process of lipid peroxidation [57]. In other words, gamma-tocopherol protects membranes from oxidation by reacting with lipid radicals. There are no free radical intermediates left after this process and the reaction comes to a complete and total stop. Oxidized-tocopheroxyl radicals are generated because of this procedure. Activated radicals can be regenerated back into the reduced form by other antioxidants as ascorbate, retinol, or ubiquinol [58, 59].

7.3 Exogenous antioxidants

Several pharmacological properties are attributed to flavonoids, a class of polyphenolic chemicals with a benzo—pyrone structure that is abundantly found in plants. Researchers have been looking into the antioxidant properties of these compounds because of the free radical scavenging and metal ion chelating abilities of their functional hydroxyl groups. Functional groups play a critical role in determining activities such as ROS/RNS-scavenging and metal chelation, which are dependent on functional groups' configuration and substitution. Flavonoid suppresses ROS creation, inhibits enzymes, or chelates trace elements that generate free radicals; Flavonoid scavenges ROS; and Flavonoid increases antioxidant protection [60].

Due to Genistein's wide range of pharmacological properties, it is perhaps the most interesting and thoroughly researched flavonoid component in soy. An abundance of research has shown that genistein can scavenge ROS and RNS with great efficiency. Through gene and protein regulation, this flavonoid molecule can increase a cell's antioxidant defenses, hence preventing apoptosis. Many plant-based foods (fruits, oils, seeds, etc.) include flavonoids, a class of naturally occurring substances that have been proved to be beneficial to human health. However, there are certain possible hazards to human health and well-being when these foods are included in the diet (as food enrichment or as nutraceuticals) [61]. Flavonoids, a group of polyphenolic compounds with a benzo—pyrone structural arrangement that is abundant in plants, are known to have several pharmacological characteristics. Because of their functional hydroxyl groups' ability to scavenge free radicals and chelate metal ions, these compounds have been investigated for their antioxidant capabilities [62].

It is important to note that the methods of antioxidant activity such as ROS/RNS removal and metal chelation are all dependent on a variety of factors. The conformational disposition of functional groups in these compounds determines their antioxidant activity. Increased antioxidant defenses and suppression of ROS generation are both caused by flavonoid-induced ROS scavenging, as are enzyme inhibition and trace element chelation [61].

In terms of pharmacological effects, genistein is perhaps the most fascinating and well-studied flavonoid molecule. It is an isoflavone found in soy. Antioxidant genistein has been used widely in a wide range of investigations, indicating its ability to scavenge ROS and RNS. Antioxidant defenses of a cell can be improved by this flavonoid molecule, which modulates numerous genes and proteins. Flavonoids are a class of naturally occurring compounds found in a wide variety of plant-based foods (fruits, oils, seeds, etc.) that have been shown to be beneficial to human health,

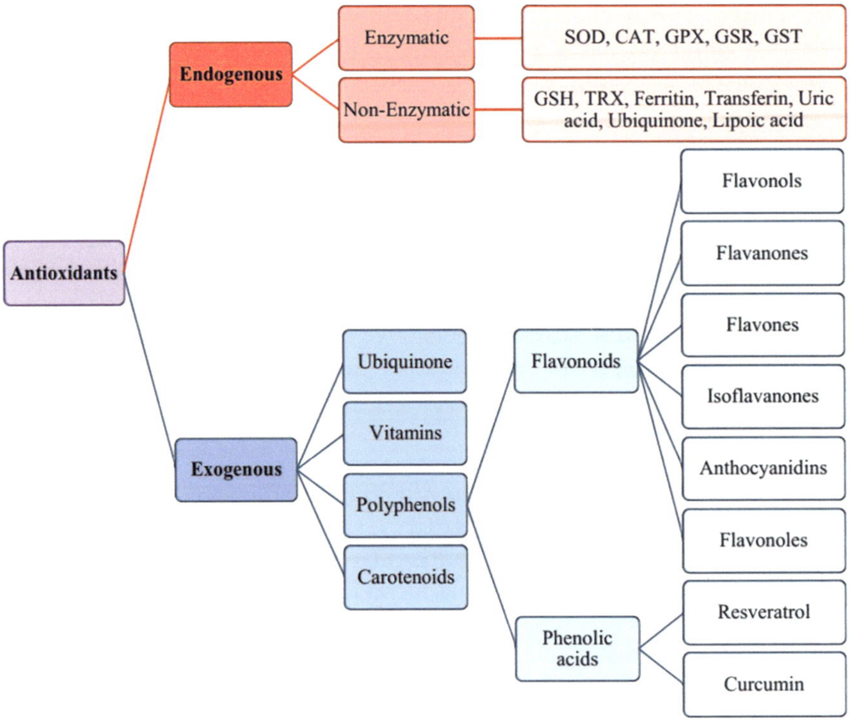

Figure 1.
Classes of antioxidants. Source: [24, 26].

both as antioxidant molecules and for their other, less obvious but no less intriguing, pharmacological qualities. It's still important to take precautions when using these supplements, and there may be some negative effects on human health and well-being if they are used in this way (as dietary supplements or nutraceuticals). Lipid soluble vitamins acts as an antioxidant, blocking free radicals from causing damage to cell membranes through a process known as lipid peroxidation (**Figure 1**) [8, 61].

8. Diets rich in antioxidants

8.1 Fruits

A lot of different fruits have a lot of health benefits, like being high in antioxidants, filled with vitamins, and having a lot of different vitamin content. Cranberries, red grapes, peaches, raspberries, strawberries, red currants, figs, cherries, pears, guava, oranges, apricots, mango, red grapes, cantaloupe, watermelon, papaya, and tomatoes are some of the fruits that fall into this category [63].

8.2 Dried fruits

Dried fruits have a greater antioxidant ratio than fresh fruits since the water have been removed during drying. They are convenient to carry around in a handbag, briefcase, or car because they can be eaten on the go and are high in nutritional value.

8.3 Vegetables

In addition to broccoli, spinach, carrots, and potatoes, other vegetables and fruits that are high in antioxidants include artichokes, cabbage, asparagus, avocados, beetroot, radish, lettuce, sweet potatoes, squash, pumpkin, collard greens, and kale. Broccoli, spinach, carrots, and potatoes are all high in antioxidants.

8.4 Herbs and other seasonings

Antioxidants can be found in a variety of spices, including cinnamon, cardamom, paprika, oregano, and turmeric. Curry powder and mustard seed powder are also rich in antioxidants. In addition to spices like sage and tarragon and herbs like peppermint and oregano and basil, herbs also include dill weed and marjoram. Dill weed is one of several herbs that can be found. They are a fantastic source of antioxidants, as well as taste and complexity to your meals.

Grains and nuts can be found in a wide variety of cuisines. Everything from cereal to nuts to a peanut butter and jelly sandwich contains nutritional powerhouses including peanut butter and jelly, granola bars, corn flakes, and granola.

8.5 Beverages

In contrast to popular belief, the great majority of our body's antioxidants can be found in beverages rather than in food. Apple juice, cider, tomato juice, pomegranate juice, and pink grapefruit juice are among the most common sources. In addition to green tea, black tea and ordinary tea contain a significant number of antioxidants. Coffee aficionados, welcome! Coffee is heavy in calories; however, it should be used in moderation because it can boost blood pressure and heart rate, which is why it is important to drink it in moderation. The antioxidants in coffee or tea are inhibited from being released when milk is added. While red wine and beer [which are both brewed from grains] provide a large amount of alcohol in moderation, the health benefits of moderate alcohol use have been extensively studied. Colorful fruits and vegetables are vital to have in your diet. Consider all selections, not just the most popular ones. More antioxidants can be found in foods that are deeper and brighter in color, such as oranges and yellows. With so many options, you'll never get bored or run out of tasty and nutritious dishes to pick from. Variety, it is claimed, is the flavor of life [64].

9. Mechanism of action of antioxidants

It has been found that antioxidants can have two primary effects. The primary antioxidant breaks the chain by supplying an electron to the system's free radical. To eliminate ROS/reactive nitrogen species initiators, the second procedure involves quenching the chain-initiating catalyst (secondary antioxidants). Co-antioxidants, electron donation, and gene expression control are some of the ways antioxidants can alter biological systems.

9.1 Preventive antioxidants

The production of ROS, such as H_2O_2 and $O_2^{\bullet}$, cannot be prevented during metabolism. To decrease the harm caused by oxidative stress, numerous techniques have been developed. One of the finest defenses against the creation of free radicals is the cell's own synthesis of antioxidant enzymes. SOD, CAT, GPx, GR, GST, thioredoxin reductase, and hemeoxygenase are some of the most significant antioxidant enzymes in the body. Using the Fenton reaction, the CAT, GPx, and CAT reactions, SOD decomposes $O_2^{\bullet}$ into water [65].

GST and GPX work together to help the body get rid of peroxides. The GSH/GSSG ratio is a well-established biomarker of oxidative stress since GRd controls it. GRd plays a vital role in boosting GSH concentration to maintain a steady oxido-redox state. As a result, researchers have discovered a link between oxidative stress and autism. Free radical generation and GSH/GSSG ratio in autistic cells were shown to be lower in comparison to those in control cells in an experiment.

Thus, GPx is widely distributed in cells, as opposed to CAT, which is typically restricted to the peroxisomes. Seven times more GPx activity in the brain than CAT activity, which is more susceptible to free radical damage, is found in the brain. Catalase (CAT) can breakdown H_2O_2 in the liver, kidneys, and erythrocytes at high doses [66].

9.2 Free radical scavengers

9.2.1 Scavenging superoxide and other ROS

The most prevalent kind of cellular free radical, which is referred to as superoxide ($O_2^{\bullet}$), is accountable for a wide array of damaging modifications. Alterations in peroxidative processes and low antioxidant concentrations are commonly related with these changes. As a means of generating more powerful $OH^{\bullet}$ and ONOO even when it itself is not reactive with biomolecules, $O_2\bullet$ is nevertheless beneficial. The phagocyte enzyme known as NADPH oxidase is responsible for the production of massive volumes of oxygen dioxide (O_2) during the process of phagocytes eliminating microorganisms. In addition to this, it is a direct result of the respiration that takes place in the mitochondria of the cell [2, 14].

9.2.2 Scavenging hydroxyl radical and other ROS

When compared to other radical species, the hydroxyl radical, which is represented by the symbol $OH^{\bullet}$ and has the chemical formula OH, is a very active radical that can cause significant damage to biological components such as DNA, lipids, and proteins. Usually, it is believed that the synthesis of $OH^{\bullet}$ originates from the Fenton reaction system. This reaction system involves the interaction of $FeSO_4$ and H_2O_2 and is then carried out in an aqueous medium. Because of this, the activity of antioxidants as $OH^{\bullet}$ scavengers can be achieved through direct scavenging, the restriction of $OH^{\bullet}$ generation through the chelation of free metal ions, or the transformation of H_2O_2 into other molecules that are not harmful to the body [9].

9.2.3 Metal ion (Fe^{2+}, Fe^{3+}, cu^{2+}, and Cu^{+}) chelating

Even though trace minerals are essential for human nutrition, they are also capable of performing the function of antioxidants (through enhancing formation of free radicals). During the process of dismutation of SOD, H_2O_2 is produced as a byproduct. This H_2O_2 combines with the ions Fe^{2+} and Cu^{+} to produce the highly reactive $OH^{\bullet}$. Although this is not the case with $OH^{\bullet}$, the reaction of iron and copper with H_2O_2 results in the generation of more singlet oxygen than $OH^•$. Oxidation has occurred for both the Fe^{2+} and the Cu^{+1} ions. When vitamin C is available, it is feasible to recycle the cellular reductants, such as NADH as well as Fe^{3+} and Cu^{2+}, to produce $OH^•$ radicals [1]. This can be accomplished when vitamin C is present. OH, is one of the most reactive elements in the body. It may react directly with proteins and fats to generate carbonyls [aldehydes and ketones], as well as cause lipid peroxidation and cross-linking. Chelating metal ions can lower their activity, which in turn lowers the formation of reactive oxygen species (ROS). Cu^{2+} and ascorbic acid are responsible for the generation of Cu^+. This Cu^+ is then chelated by the Se antioxidant, which prevents DNA damage caused by the $OH^•$ radical that is generated when Cu^+ is combined with H_2O_2 [67].

9.3 Free radical generating enzyme inhibitors

The production of free radicals by specific enzymes has been shown to occur in a wide range of physiological and pathological situations. The plasma membrane is home to a group of enzymes known as NADPH oxidases. A cytosolic donor NADPH is transferred to an extracellular oxygen molecule by these electron transporters. Hypoxanthine and hypoxanthin are converted into uric acid in the organism when catalysts for the oxidation of these compounds are present [2]. Hydrogen peroxide and oxygen radicals are formed because of this process. Aside from mitochondrial respiration, additional enzymes produce oxygen dioxide as a waste product, including NADH oxidase, monooxygenases, and cyclooxygenases. The enzyme NADPH oxidase produces a large amount of oxygen that is poisonous to all living things to fight infections in a way that is dependent on oxygen. The dying mechanism then makes use of this oxygen. Regulating the generation of reactive oxygen derivatives is critical during a respiratory burst to prevent tissue damage. This is done so that an organism can defend itself against invading pathogens [68]. On the other side, excessive ROS can cause oxidative stress, which can lead to processes like the oxidation of low-density lipoprotein (LDL). An increase in the amount of oxidized LDL that is circulating in the blood of patients who have metabolic syndrome has been linked to increased phagocytic NADPH oxidase activity. To lessen the negative consequences of oxidative stress, several studies have demonstrated that hemodialysis patients can gain advantages from inhibiting NADPH oxidase and taking antioxidants. In recent years, a great number of naturally occurring antioxidants have demonstrated the potential to inhibit enzymes that enhance O_2 generation, which has led to the development of new therapy agents for illnesses related to oxidative stress [61].

9.4 Prevention of lipid peroxidation

Some of the most frequent C-C double bonds can be found in unsaturated fatty acids, glycolipids, cholesterol, the cholesterol ester, and phospholipids, although there are many others. Lipid peroxidation is the process by which these compounds are oxidized. As a result of this chain reaction, ROS begin to damage unsaturated fats. There are several double bonds and methylene-CH_2-group groups in unsaturated fatty acids, which are very reactive hydrogens. Antioxidants can quench peroxide radicals directly, stopping the chain reaction and preventing further damage [1]. Chronic diseases such as cancer and atherosclerosis, which can cause early death, have been related to lipid peroxidation. It is possible for antioxidant compounds to neutralize or inhibit the generation of ROS and peroxide radicals, respectively. Lipid peroxidation is an important method for discovering naturally occurring antioxidants and figuring out the mechanism by which they work. Lipoproteins and red blood cells as well as low density lipoprotein (LDL) have been studied for their ability to protect against lipid peroxidation due to free radicals. An important aspect influencing the antioxidant activity of these polyphenols is their structure and starting circumstances. The microenvironment in which this reaction occurs also has a substantial impact [61].

9.5 Prevention of DNA damage

Direct interactions between nitric oxide radicals ($OH^{\bullet}$) and $O_2^{\bullet}$ radicals ($OO^{\bullet}$) in living cells can sever a single strand of DNA. This causes damage to the DNA that is not repairable in any way. DNA damage, which results in cell death and mutation, has been linked to degenerative diseases such as Alzheimer's. As a result of this, DNA or plasmid damage has evolved into a model for the investigation and characterization of antioxidants. On the other hand, a study that used metal-free plasmid DNA found that the reaction of Cu^{2+} with ascorbic acid and hydrogen peroxide at pH 7 caused DNA damage. This was observed in the study. During the process, which takes place in the presence of Cu^{2+}, ascorbic acid is utilized to bring about the reduction of Cu^{2+} to Cu^{+}. Cu^{+} and H_2O_2 react to produce an $OH^{\bullet}$ radical, which then cleaves one strand of DNA. This results in the typically supercoiled plasmid DNA unraveling and becoming more helical [69].

10. Conclusion

Oxidative stress occurs in cells as a result of normal physiological processes and interactions with the environment, and cells are protected from oxidative damage by a sophisticated network of antioxidant defense systems. In general, our body's innate antioxidant defense system or antioxidants added to our diet counteract the creation of reactive species. Oxidative stress arises when this balance is disrupted. Oxidative stress plays a role in the development of a wide range of diseases, including those of the gastrointestinal system. The development of antioxidant therapies is a viable route for the treatment of gastrointestinal illnesses, as data suggests that using antioxidants can improve the progression of many diseases. As a result, understanding the unique oxidative route implicated in each disease may help to identify disease signs as well as design preventive and curative therapy techniques. Inhibiting radical generation, scavenging radicals, or stimulating their breakdown are some of the ways antioxidants protect tissue against free radical harm. In the last few years, synthetic antioxidants have been connected to human health issues. As a result, the search for natural antioxidative compounds that are safe and effective has intensified in recent years. Dietary and plant-derived antioxidants may offer a feasible alternative to the body's own antioxidant defenses. A wide variety of plant and food-derived antioxidants can be found.

Conflict of interest

The authors do not have any conflict of interest to declare.

Author details

Olalekan Bukunmi Ogunro[1*], Aderonke Elizabeth Fakayode[2] and Gaber El-Saber Batiha[3]

1 Reproductive and Endocrinology, Toxicology, and Bioinformatics Research Laboratory, Department of Biological Sciences, KolaDaisi University, Ibadan, Nigeria

2 Department of Biochemistry, Federal University of Technology, Akure, Nigeria

3 Faculty of Veterinary Medicine, Department of Pharmacology and Therapeutics, Damanhour University, AlBeheira, Egypt

*Address all correspondence to: olalekanbukunmi@gmail.com

References

[1] Lobo V, Patil A, Phatak A, Chandra N. Free radicals, antioxidants and functional foods: Impact on human health. Pharmacognosy Reviews. 2010;**4**(8):118-126

[2] Kurutas EB. The importance of antioxidants which play the role in cellular response against oxidative/nitrosative stress: Current state. Nutrition Journal. 2016;**15**(1):71

[3] Tan BL, Norhaizan ME, Liew WPP, Rahman HS. Antioxidant and oxidative stress: A mutual interplay in age-related diseases. Frontiers in Pharmacology. 2018;**9**(OCT):1-28

[4] Sharifi-Rad M, Anil Kumar NV, Zucca P, Varoni EM, Dini L, Panzarini E, et al. Lifestyle, oxidative stress, and antioxidants: Back and forth in the pathophysiology of chronic diseases. Frontiers in Physiology. 2020;**11**(July):1-21

[5] Conti V, Izzo V, Corbi G, Russomanno G, Manzo V, De Lise F, et al. Antioxidant supplementation in the treatment of aging-associated diseases. Frontiers in Pharmacology. 2016;7(FEB):1-11

[6] Azat Aziz M, Shehab Diab A, Abdulrazak MA. Antioxidant categories and mode of action. In: Antioxidants. London, United Kingdom: IntechOpen; 2019. DOI: 10.5772/intechopen.83544

[7] Lourenço SC, Moldão-Martins M, Alves VD. Antioxidants of natural plant origins: From sources to food industry applications. Molecules. 2019;**24**(22):14-16

[8] Martemucci G, Costagliola C, Mariano M, D'andrea L, Napolitano P, D'Alessandro AG. Free radical properties, source and targets, antioxidant consumption and health. Oxygen. 2022;**2**(2):48-78

[9] Phaniendra A, Jestadi DB, Periyasamy L. Free radicals: Properties, sources, targets, and their implication in various diseases. Indian Journal of Clinical Biochemistry. 2015;**30**(1):11-26

[10] Buettner GR, Schafer FQ. Free radicals, oxidants, and antioxidants. Teratology. 2000;**62**(4):234

[11] Demarquoy J. Crosstalk between mitochondria and peroxisomes. World Journal of Biological Chemistry. 2015;**6**(4):301

[12] Kumari N, Dwarakanath BS, Das A, Bhatt AN. Role of interleukin-6 in cancer progression and therapeutic resistance. Tumor Biology. 2016;**37**(9):11553-11572. DOI: 10.1007/s13277-016-5098-7

[13] Li YR, Trush M. Defining ROS in biology and medicine. Reactive Oxygen Species. 2016;**1**(1):9-21

[14] Pham-Huy LA, He H, Pham-Huy C. Free radicals, antioxidants in disease and health. International Journal of Biomedical Sciences. 2008;**4**(2):89-96

[15] Singh R, Devi S, Gollen R. Role of free radical in atherosclerosis, diabetes and dyslipidaemia: Larger-than-life. Diabetes/Metabolism Research and Reviews. 2015;**31**(2):113-126

[16] Cadet J, Davies KJA. Oxidative DNA damage & repair: An introduction. Free Radical Biology & Medicine. 2017;**107**:2-12

[17] Valavanidis A, Vlachogianni T, Fiotakis C. 8-Hydroxy-2′ -

deoxyguanosine (8-OHdG): A critical biomarker of oxidative stress and carcinogenesis. Journal of Environmental Science and Health, Part C: Environmental Carcinogenesis and Ecotoxicology Reviews. 2009;**27**(2):120-139

[18] Paiva CN, Bozza MT. Are reactive oxygen species always detrimental to pathogens? Antioxidants Redox Signal. 2014;**20**(6):1000-1034

[19] Di Meo S, Reed TT, Venditti P, Victor VM. Role of ROS and RNS sources in physiological and pathological conditions. Oxidative Medicine and Cellular Longevity. 2016;**2016**:1245049

[20] Levine B, Kroemer G. Autophagy in the pathogenesis of disease. Cell. 2008;**132**(1):27-42

[21] Bonam SR, Wang F, Muller S. Lysosomes as a therapeutic target. Nature Reviews. Drug Discovery. 2019;**18**(12):923-948. DOI: 10.1038/s41573-019-0036-1

[22] Sánchez-álvarez M, Strippoli R, Donadelli M, Bazhin A V., Cordani M. Sestrins as a therapeutic bridge between ros and autophagy in cancer. Cancers (Basel). 2019;**11**(10):1415

[23] Wible DJ, Bratton SB. Reciprocity in ROS and autophagic signaling. Current Opinion in Toxicology. 2018;7:28-36. DOI: 10.1016/j.cotox.2017.10.006

[24] Nita M, Grzybowski A. The role of the reactive oxygen species and oxidative stress in the Pathomechanism of the age-related ocular diseases and other pathologies of the anterior and posterior eye segments in adults. Oxidative Medicine and Cellular Longevity. 2016;**2016**:3164734

[25] Tsai MF, Chen SM, Ong AZ, Chung YH, Chen PN, Hsieh YH, et al. Shikonin induced program cell death through generation of reactive oxygen species in renal cancer cells. Antioxidants. 2021;**10**(11):1831

[26] Liu H, Nishitoh H, Ichijo H, Kyriakis JM. Activation of apoptosis signal-regulating kinase 1 (ASK1) by tumor necrosis factor receptor-associated factor 2 requires prior dissociation of the ASK1 inhibitor Thioredoxin. Molecular and Cellular Biology. 2000;**20**(6):2198-2208

[27] Rodrigo R, Retamal C, Schupper D, Vergara-Hernández D, Saha S, Profumo E, et al. Antioxidant Cardioprotection against reperfusion injury: Potential therapeutic roles of resveratrol and quercetin. Molecules. 2022;**27**(8):1-23

[28] Moriwaki K, Chan FKM. RIP3: A molecular switch for necrosis and inflammation. Genes & Development. 2013;**27**(15):1640-1649

[29] Galluzzi L, Vitale I, Aaronson SA, Abrams JM, Adam D, Agostinis P, et al. Molecular mechanisms of cell death: Recommendations of the nomenclature committee on cell death 2018. Cell Death and Differentiation. 2018;**25**(3):486-541

[30] Zhang Y, Su SS, Zhao S, Yang Z, Zhong CQ, Chen X, et al. RIP1 autophosphorylation is promoted by mitochondrial ROS and is essential for RIP3 recruitment into necrosome. Nature Communications. 2016;**2017**(8):1-14. DOI: 10.1038/ncomms14329

[31] Cai Z, Liu ZG. Execution of RIPK3-regulated necrosis. Molecular & Cellular Oncology. 2014;**1**(2):1-7

[32] Rafieian-Kopaei M, Setorki M, Doudi M, Baradaran A, Nasri H. Atherosclerosis: Process, indicators, risk factors and new hopes. International

Journal of Preventive Medicine. 2014;**5**(8):927-946

[33] Hadi HAR, Al Suwaidi JA. Endothelial dysfunction in diabetes mellitus. Vascular Health and Risk Management. 2007;**3**(6):853-876

[34] Steven S, Frenis K, Oelze M, Kalinovic S, Kuntic M, Jimenez MTB, et al. Vascular inflammation and oxidative stress: Major triggers for cardiovascular disease. Oxidative Medicine and Cellular Longevity. 2019;**2019**:7092151

[35] Virdis A, Duranti E, Taddei S. Oxidative stress and vascular damage in hypertension: Role of angiotensin ii. International Journal of Hypertension. 2011;**2011**:916310

[36] Pai WY, Lo WY, Hsu T, Peng CT, Wang HJ. Angiotensin-(1-7) inhibits thrombin-induced endothelial phenotypic changes and reactive oxygen species production via NADPH oxidase 5 downregulation. Frontiers in Physiology. 2017;**8**(DEC):994

[37] García-Sánchez A, Miranda-Díaz AG, Cardona-Muñoz EG. The role of oxidative stress in physiopathology and pharmacological treatment with pro- and antioxidant properties in chronic diseases. Oxidative Medicine and Cellular Longevity. 2020;**2020**:2082145

[38] Matough FA, Budin SB, Hamid ZA, Alwahaibi N, Mohamed J. The role of oxidative stress and antioxidants in diabetic complications. Sultan Qaboos University Medical Journal. 2012;**12**(1):556-569

[39] Giacco F, Brownlee M. Oxidative stress and diabetic complications. Circulation Research. 2010;**107**(9): 1058-1070

[40] Suárez-Rivero J, Villanueva-Paz M, de la Cruz-Ojeda P, de la Mata M, Cotán D, Oropesa-Ávila M, et al. Mitochondrial dynamics in mitochondrial diseases. Diseases. 2016;**5**(1):1

[41] Singh A, Kukreti R, Saso L, Kukreti S. Oxidative stress: A key modulator in neurodegenerative diseases. Molecules. 2019;**24**(8):1-20

[42] Agnihotri A, Aruoma OI. Alzheimer's disease and Parkinson's disease: A nutritional toxicology perspective of the impact of oxidative stress, mitochondrial dysfunction, nutrigenomics and environmental chemicals. Journal of the American College of Nutrition. 2020;**39**(1):16-27. DOI: 10.1080/07315724.2019.1683379

[43] Lee Mosley R, Benner EJ, Kadiu I, Thomas M, Boska MD, Hasan K, et al. Neuroinflammation, oxidative stress, and the pathogenesis of Parkinson's disease. Clinical Neuroscience Research. 2006;**6**(5):261-281

[44] Liou GY, Storz P. Reactive oxygen species in cancer. Free Radical Research. 2010;**44**:479-496

[45] Stancill JS, Broniowska KA, Oleson BJ, Naatz A, Corbett JA. Pancreatic -cells detoxify H2O2 through the peroxiredoxin/thioredoxin antioxidant system. The Journal of Biological Chemistry. 2019;**294**(13):4843-4853

[46] Wang Y, Branicky R, Noë A, Hekimi S. Superoxide dismutases: Dual roles in controlling ROS damage and regulating ROS signaling. The Journal of Cell Biology. 2018;**217**(6):1915-1928

[47] Zafra A, Castro AJ, de Alché JD. Identification of novel superoxide dismutase isoenzymes in the olive (Olea

europaea L.) pollen. BMC Plant Biology. 2018;**18**(1):1-16

[48] Folgueira I, Lamas J, de Felipe AP, Sueiro RA, Leiro JM. Identification and molecular characterization of superoxide Dismutases isolated from a Scuticociliate parasite: Physiological role in oxidative stress. Scientific Reports. 2019;**9**(1):1-14

[49] Hassan EA, Al-Zuhairi WS, Ibrahim WA. Antioxidants and their role in preventing diseases: A review. Earthline Journal of Mathematical Sciences. 2022;7(2):165-182

[50] Zamocky M, Furtmüller PG, Obinger C. Evolution of catalases from bacteria to humans. Antioxidants Redox Signal. 2008;**10**(9):1527-1547

[51] Traber MG, Stevens JF. Vitamins C and E: Beneficial effects from a mechanistic perspective. Free Radical Biology & Medicine. 2011;**51**(5):1000-1013

[52] Kasote DM, Katyare SS, Hegde MV, Bae H. Significance of antioxidant potential of plants and its relevance to therapeutic applications. International Journal of Biological Sciences. 2015;**11**(8):982-991

[53] Lushchak VI. Glutathione homeostasis and functions: Potential targets for medical interventions. Journal of Amino Acids. 2012;**2012**:1-26

[54] Rehman T, Shabbir MA, Inam-Ur-Raheem M, Manzoor MF, Ahmad N, Liu ZW, et al. Cysteine and homocysteine as biomarker of various diseases. Food Science & Nutrition. 2020;**8**(9):4696-4707

[55] Couto N, Wood J, Barber J. The role of glutathione reductase and related enzymes on cellular redox homoeostasis network. Free Radical Biology & Medicine. 2016;**95**:27-42. DOI: 10.1016/j.freeradbiomed.2016.02.028

[56] Azzi A. Many tocopherols, one vitamin E. Molecular Aspects of Medicine. 2018;**61**:92-103. DOI: 10.1016/j.mam.2017.06.004

[57] Singh VK, Beattie LA, Seed TM. Vitamin E: Tocopherols and tocotrienols as potential radiation countermeasures. Journal of Radiation Research. 2013;**54**(6):973-988

[58] Domej W, Oettl K, Renner W. Oxidative stress and free radicals in COPD-implications and relevance for treatment. International Journal of COPD. 2014;**9**:1207-1224

[59] Silva SAM, Michniak-Kohn B, Leonardi GR. An overview about oxidation in clinical practice of skin aging. Anais Brasileiros de Dermatologia. 2017;**92**(3):367-374

[60] Rudrapal M, Khairnar SJ, Khan J, Bin DA, Ansari MA, Alomary MN, et al. Dietary polyphenols and their role in oxidative stress-induced human diseases: Insights into protective effects, Antioxidant Potentials and Mechanism(s) of Action. Frontiers in Pharmacology. 2022;**13**(February):1-15

[61] Pizzino G, Irrera N, Cucinotta M, Pallio G, Mannino F, Arcoraci V, et al. Oxidative stress: Harms and benefits for human health. Oxidative Medicine and Cellular Longevity. 2017;**2017**:8416763

[62] Panche AN, Diwan AD, Chandra SR. Flavonoids: An overview. Journal of Nutrition Sciences. 2016;**5**:e47

[63] Vinson JA, Zubik L, Bose P, Samman N, Proch J. Dried fruits: Excellent in vitro and in vivo antioxidants. Journal of the American College of Nutrition. 2005;**24**(1):44-50

[64] Seeram NP, Aviram M, Zhang Y, Henning SM, Feng L, Dreher M, et al. Comparison of antioxidant potency of commonly consumed polyphenol-rich beverages in the United States. Journal of Agricultural and Food Chemistry. 2008;**56**(4):1415-1422

[65] Lubos E, Loscalzo J, Handy DE. Glutathione peroxidase-1 in health and disease: From molecular mechanisms to therapeutic opportunities. Antioxidants Redox Signal. 2011;**15**(7):1957-1997

[66] Dumanović J, Nepovimova E, Natić M, Kuča K, Jaćević V. The significance of reactive oxygen species and antioxidant defense system in plants: A concise overview. Frontiers in Plant Science. 2021;**11**(January):552969

[67] Lü JM, Lin PH, Yao Q, Chen C. Chemical and molecular mechanisms of antioxidants: Experimental approaches and model systems. Journal of Cellular and Molecular Medicine. 2010;**14**(4):840-860

[68] Bhattacharyya A, Chattopadhyay R, Mitra S, Crowe SE. Oxidative stress: An essential factor in the pathogenesis of gastrointestinal mucosal diseases. Physiological Reviews. 2014;**94**(2):329-354

[69] Islam B, Habib S, Ahmad P, Allarakha S, Moinuddin Ali A. Pathophysiological role of peroxynitrite induced DNA damage in human diseases: A special focus on poly(ADP-ribose) polymerase (PARP). Indian Journal of Clinical Biochemistry. 2015;**30**(4):368-385. DOI: 10.1007/s12291-014-0475-8